How To Use Elements Effectively

Author:

Trevor Hellen

ACKNOWLEDGEMENT

I wish to take this opportunity as Chairman of NAFEMS Education and Training Working Group to thank the Working Group members for their help and support in the preparation of this booklet. The composition of the Working Group is:

Adib Becker	University of Nottingham
Dave Ellis	IDAC
Trevor Hellen	Consultant
Bob Johnson	DAMT Ltd
Dermot Monaghan	Discrete Simulations Ltd
Derek Pashley	Consultant
Anup Puri	BAE Systems
Darius Sepahy	Consultant
John Smart	North East Wales Institute, Wrexham
Bryan Spooner	Consultant
Jim Wood	University of Strathclyde

In particular, I would like to thank Bryan Spooner in his capacity as chief reviewer of this booklet, and to Bob Johnson for help in providing a number of the figures and results.

John Smart
Chairman

Preface

NAFEMS is a non-profit making association of organisations using, developing, or teaching the various forms of numerical analysis in common practice.

This booklet is a continuation of the "How and Why…" set of booklets published by NAFEMS, designed to guide both new and experienced analysts in a range of problem types. The booklets are written to introduce various analysis methodologies to both engineering managers and engineers, in a straightforward and informed manner. They are complemented by more detailed publications from NAFEMS; for example, the *Finite Element Primer*.

Linear finite element analysis has for many years been widely used in the civil and mechanical engineering fields and, in particular, in the construction, automotive, aerospace, nuclear and offshore sectors. Such analysis is an integral part of the design cycle in many companies. Finite element programs that have the capability to solve non-linear problems have also been available for many years. Initially used in the more specialised industries typified by nuclear and aerospace engineering, non-linear finite element analysis is now applied to nearly all areas of engineering, using commercially available packages of high quality and reliability. However, all this software relies on the good use of elements and suitable meshes.

The main aim of this How To booklet is to explain the issues involved in designing suitable meshes and selecting appropriate elements for solving such problems. The emphasis is on using the more popular types of element in elastic conditions, although the techniques and mechanics of actual mesh generation software are not covered.

Good use of elements is central to practical finite element application, but coverage of this subject is sparse compared to that of the underlying theories of the method. In fact, it is a considerably more nebulous subject and more complicated to explain. Hence, the present work, based on the author's personal experience, is somewhat subjective and does not profess to establish any definitive format for this subject.

The discussions and presentation of finite element theory involved are aimed at the level of the graduate in engineering or a related discipline, who is one year into a professional engineering career, but with a wide audience of practising and potential finite element users also in mind. Some basic knowledge of the finite element method is assumed.

Contents

1. Introduction

The main aim of this How To booklet is to explain the issues involved in designing suitable meshes and selecting appropriate elements for solving such problems. The emphasis is on using the more popular types of element in elastic conditions, although the techniques and mechanics of actual mesh generation software are not covered.

The use of the finite element method to solve real engineering problems is nowadays very well known, and a great many successful applications have been reported, leading to safer and more economic products. The method is often used as a backup to, or a necessary ingredient of, engineering codes of practice. Its success is due to several factors. It is available on computers of all sizes, including modern desktop units and laptops. It is a versatile and much cheaper alternative (or complement) to conducting routine experiments, being capable of solving a wide range of structural analysis and fluid flow problems under appropriate loading.

The finite element method requires a model of the actual structure to be defined as a geometric mesh of notional grid lines on and through the structure, along with definitions of boundary conditions, loads, material properties, and other details. The mesh defines an assemblage of elements, each capable of some prescribed but relatively simple field behaviour. The more elements, the more accurate the results, satisfying a basic property of the method, that, generally, the solution converges to the correct one with increasing numbers of nodes (or elements). However, the cost of analysis increases as well, both in terms of computer time and memory and disc requirements. A large 3D job can soon consume all available computer resources. For example, if the elements in a 3D mesh are doubled in each direction, so that every element is replaced by eight new ones, the direct solution time increases by several orders of magnitude, all other things being equal. Hence a compromise mesh in terms of numbers of elements, to be accurate enough whilst maintaining reasonable demands on computer resources, has to be found.

The calculated results are primary and secondary field variables at each reference point in the structure, usually at nodes or internal points in elements. The primary variables are the degrees of freedom, which are derived by solving the equilibrium equations, and are, typically, displacements in stress and dynamics solutions, temperatures in thermal cases, or pressures in fluids. The secondary variables are derived from the primaries, typified by stresses, heat fluxes, etc., and are pointwise geometric derivatives of the primary variables, solved *a posteriori* in each element from the primary variables.

The distinction between primary and secondary variables in this booklet is important, since these are the key products of the finite element solution, and the errors in the calculations are deduced from these variables. In particular, secondary variables are useful indicators of inadequate mesh refinement when using isoparametric elements, which are by far the most common element types in use today. Inadequate mesh refinement is usually due to unfamiliarity with the problem and not knowing what an adequate mesh should look like. It can also be the product of an inexperienced user, possibly trying to use an insufficient number of elements, or using badly shaped elements. Secondary variables are also used to drive automatic self-adaptive meshing programs, where an iteration process performs a required analysis, checks the results to derive secondary variable error indicators, automatically redesigns the mesh based on these indicators, then returns to the start of the loop for another analysis.

A dominant activity when using the finite element method is mesh generation. Although the results can be very valuable and the most complicated 2D and 3D shapes, with many thousands of degrees of freedom, can now be solved, still the most time consuming aspect for the user often lies in preparing the mesh. To help this, many mesh generation programs have been written, and are regularly in successful use. However, as problems become more complicated, even these programs become stretched and difficult to use. In practice, automatic self-adaptive meshing programs are not yet sufficiently versatile to cover all meshing needs, in particular for complicated 3D shapes where they would be particularly useful. Even there, though, care and human intervention may be required to ensure meshes generated by such schemes do not exceed computer limits.

Because of the nature of the mathematical equations of physics that are being solved, some areas of the mesh require relatively small elements where primary and secondary variables are changing rapidly with respect to geometry, whereas where these changes are slower, larger elements will suffice. The perfect mesh design will strive to produce a uniform discretisation error throughout the mesh. Keeping this distribution of element sizes, successive levels of further refinement will decrease this error. In some very simple problems, it is possible to achieve a 0% error, although in practical cases all that is required is to achieve a working error, of a given percentage, depending on the importance and type of the results (primary variables are calculated more accurately than secondary variables). Unfortunately, the finite element method cannot furnish a single number for this discretisation error, so again some user skill in mesh usage and results interpretation is important.

Another problem which arises in mesh design is the over-distortion of element shapes, known as **shape sensitivity**. Elements are designed assuming a relatively simple field behaviour that is defined with respect to element shape. In 2D, for instance, most elements are either triangular or quadrilateral, with nodes at the vertices and also possibly along the sides. The basic reference shapes are, respectively, equilateral triangles and squares, and any change to these shapes in actual usage is known as a distortion. This can occur near curved geometric features such as boundaries or holes, or when going from areas of high to low mesh refinement. The mesh designer should be aware that such distortions can lead to additional solution errors, although most element families, in particular the isoparametric elements, can tolerate considerable distortion with no adverse effects. Indeed, in some cases certain distorted shapes enhance accuracy, as described in chapter 5. Some quantifying measures are available and described in this booklet.

The above comments help to illustrate that creating meshes that are efficient and fit for purpose can be a complicated exercise. However, every user of finite elements has to produce a model, with mesh, boundary conditions, loads, etc., and so needs to be very aware of these issues. Unfortunately, most practising engineers have to achieve this, often with little previous relevant training, and in an urgent time scale that prohibits any chance of formal learning. They will probably be using commercial finite element mesh generation software, which itself requires an education and training schedule before it can be used to solve routine problems with confidence and success. Although some learning time will therefore have to be made available in the engineer's busy schedule, it is hoped that this booklet and some of the references herein will help to provide suitable knowledge in a speedy and efficient manner. It is most important that a finite element user develops a good instinct and culture of model preparation and, in particular, mesh design. These comments apply even if all meshing can be done using a pre-processor in a fairly automatic way.

The finite element method is the most versatile numerical method used in solving the different forms of the mathematical equations of physics. Other numerical methods exist, such as:

- semi-analytical; mathematical solutions, which are of very limited scope,
- boundary element method,
- finite difference method.

The boundary element method has some useful advantages, since only the boundary has to be modelled and therefore one less dimension of meshing suffices compared to finite elements. However, the method is much less versatile than finite elements for most types of non-linear analysis, and there is only a limited amount of commercial software available. Another promising development within the finite element method is that of "meshless" meshes, where only nodes are defined, and the software creates its own integration points, without the need to describe individual elements.

This booklet concentrates on finite elements, for which reference will be made to linear static elastic stress analysis in particular. Here, when the displacement formulation is used, the primary variables (the main degrees of freedom) are generalised displacements, and the secondary variables are stresses and strains.

Many of the developments will be directly applicable to extensions to linear static elastic analysis, such as material and non-linear behaviour and time-dependent effects such as dynamics. Descriptions will refer mainly to the isoparametric element family, the most general and useful in practice, applicable to 2D plane, 2D axisymmetric, and 3D geometries. Beams, plates and shell elements are not covered explicitly, although a lot of the principles will be relevant. Note that the generalised displacements include displacements, rotations, and further derivative terms for beams, plates and shells.

NAFEMS has produced several other "How To…" and "Why Do…" booklets in recent years on themes related to the practical aspects of meshing. These include:

- Why Do Finite Element Analysis [1],
- How To Plan a Finite Element Analysis [2],
- How To Get Started with Finite Elements [3],
- How To Choose a Finite Element Pre- and Post-Processor [4],
- How To Model with Finite Elements [5],
- Tips and Workabouts for CAD Generated Models [6],
- How To Interpret Finite Element Results [7].

These texts offer good basic material related to finite element modelling. The present booklet is complementary to these and concentrates on its stated aims of the production of appropriate meshes for solving practical problems, with emphasis on using the chosen types of element in an effective and proper manner.

The open literature refers almost exclusively to the more theoretical aspects of finite elements and their use in application. Of the many text books available, the two which have been consulted in writing this booklet are by Cook et al [8] and Zienkiewicz and Taylor [9]. Another useful text dedicated to the practical user of the method is by Adams and Askenazi [10], whilst MacNeal [11] concentrates on the design and performance of finite element families.

2. The Main Types of Element

2.1 Introduction

All finite element models use meshes of elements of type(s) depending on the structure. In reality, all structures are 3-dimensional, so that 3D elements are the most general, incorporating a complete set of field variables at every point of reference. In practice, however, structures can often be represented as 2D plane, when a constant thickness exists, or 2D axisymmetric for a body of revolution. These give considerable savings in the number of field variables generated in the model, as well as both manpower and computer time. Beams, plates and shells can be modelled with similar gains in efficiency.

The **element type** chosen for a particular mesh reflects the type of structure. It should be consistent over the mesh. The element type is defined as containing a certain number of nodes, with specific primary variables at each node. Throughout the element, values of these variables are defined using prescribed **shape functions**, or interpolation functions, usually polynomials. The most common such variable is displacement, as used in stress and dynamic analyses. Thus, a linear element permits a linear variation in each displacement component (or degree of freedom, 2 in 2D, 3 in 3D), and is often termed a **linear displacement element**, or just a **linear element** (figure 2.1). **Quadratic elements** are more popular, producing greater accuracy for a given total number of variables in the mesh. **Cubic elements** are even more accurate, but are not so popular for several reasons, including clustering of nodes along the element sides. The shape functions apply equally in every dimensional direction. The above comments therefore apply to 2D and 3D. In principle, they apply also to beams, plates and shells, although here semi-analytical assumptions are used instead of the shape functions in some of the directions.

The secondary field variables are defined as derivatives of the primary variables. Thus, when displacements are primaries, strains and stresses are secondaries. Since the derivatives are with respect to spatial direction, it can be seen that the secondary variables are polynomials one order less than the primaries in 1D. In 2D and 3D, the use of mixed dimensional shape functions complicates the situation slightly, in that the polynomial representation of primaries and secondaries are now of the same order. Thus, for quadratic displacement element types, the stresses are quasi-linear rather than linear, in that some quadratic terms are present. For thermal problems, temperature and flux are the primary and secondary variables, respectively.

The **element shape** is another property of an element type, indicating a set number of nodes and sides or faces. Some elements also require definition of extra geometric features, such as thickness and curvature. In 2D, element shapes are either lines, triangles or quadrilaterals, which can represent plane, axisymmetric bodies of revolution, beam, plate and shell geometries. In 3D, they can take a range of shapes, such as tetrahedra, pyramids, wedges and hexahedra (bricks), all of which have faces which are triangular, quadrilateral or sometimes just points. In all cases, nodes are defined at specific locations depending on the element type.

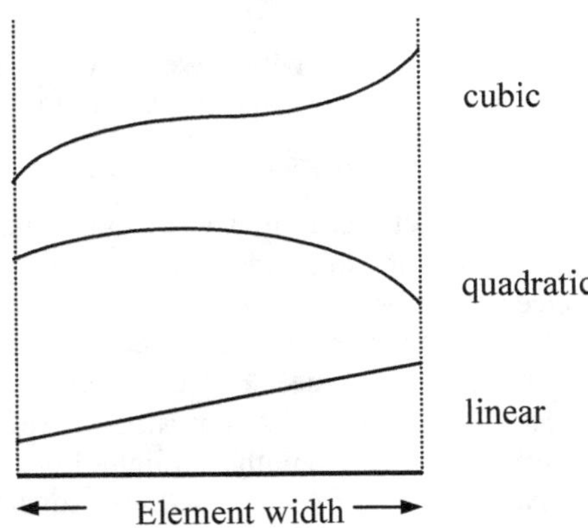

Figure 2.1 Linear, Quadratic and Cubic Variations across an Element

An important rule in the choice of elements is that mixtures of element types in the same mesh should be made with care. Thus, mixtures of 2D, 3D, beam, plate and shell elements should only be made when physically admissible. The mixing of elements with different orders of polynomial is theoretically inadmissible. This means that, if intermediate nodes exist along a common side or face in one element, then corresponding nodes should also exist in other elements sharing that side or face. Otherwise, along that common side or face, the elements will not be **compatible**, or **conforming**. This is an important property that is required in most element types, although a particular incompatible type of element is described in chapter 3. Different shaped elements may be used together in the same mesh as long as they conform along their common sides or faces. Lack of conformity can be controlled in some software by subtle use of multiple point constraints.

This chapter describes some of the more important element types, including those theoretical aspects that are relevant to providing an understanding of how elements behave in practice. More comprehensive theoretical details may be found in finite element text books, such as [8] and [9], whilst [11] concentrates on the design and performance of finite element families.

2.2 The Main Element Families

The shape functions that define the primary field variables (the displacement component variations) can also be used to define the geometric shape of the element. This defines the well-known **isoparametric element**. It applies to 2D, 3D, beams, plates and shells. These elements are by far the most popular elements in everyday use. Closely related families are the sub-parametric and super-parametric elements. Here, we concentrate on the isoparametric family since many of the concepts apply to those others as well. They contain the constant stress triangle and the 4-noded quadrilateral, which were extensively used in the early days of FEM, plus the more modern and versatile midside node (quadratic displacement) versions.

A requirement of the shape functions is that each individual element type has a prescribed number of nodes on or along each of its sides (or edges and faces in 3D). For linear displacement elements, nodes exist only at the corners. For quadratic elements, nodes are at corners and one node along each side, nominally at the mid-point position (midside node). For cubic elements, the nodes are at corners, with two between at the one-third side positions. Moving these non-corner nodes into suitable positions, away from the straight line, defines quadratic (as parabolic) and cubic curved sides. An immediate advantage is that curved boundaries can be modelled accurately. Nodes may also exist inside the element or within faces in 3D.

The concepts and properties of isoparametric elements can most easily be seen by considering the 2D continuum subset. Either a constant thickness is assumed, for plane stress or plane strain, or there is axisymmetry, with non-zero out-of-plane strains and stresses present. Several families can be identified, each having their linear, quadratic and cubic displacement elements shown in the respective figures.

2.2.1 Serendipity Quadrilaterals

This family contains elements which are usually linear, quadratic or cubic, and whose nodes lie only on the boundary (figure 2.2). Higher orders can also be used, but are rare and require some internal nodes. The particular shape functions required were initially derived using some experimentation and ingenuity, which gave rise to the term **serendipity**, after the Princes of Serendip, noted for their chance discoveries.

For clarity in this booklet, the element types are given a name of the form *shape* plus *number of nodes,* as shown in figure 2.2. Thus, the linear quadrilateral has four nodes so is called QUAD4. These names are not to be confused with those of any particular commercial software. QUAD4 and QUAD8 are two of the most common elements in practical usage. These elements use incomplete polynomials for each order, which ironically enough can give enhanced behaviour, as discussed later.

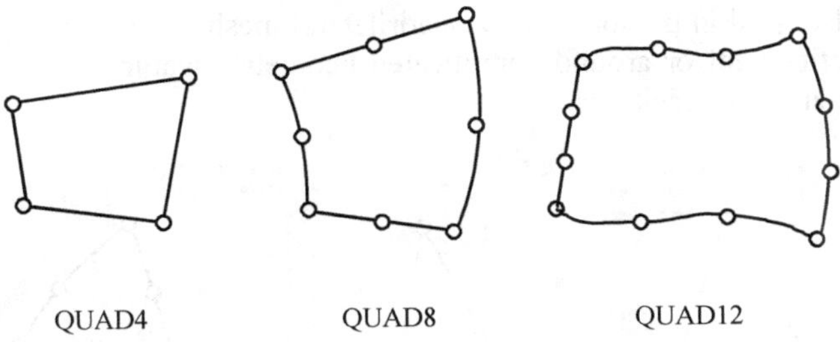

Figure 2.2 2D Serendipity Element Family

2.2.2 Lagrangian Quadrilaterals

These are similar to the serendipity family, except the shape functions are based on Lagrange polynomials, which mean that nodes are required insides the element at positions corresponding to those on boundaries (figure 2.3). The linear element QUAD4 is also a serendipity element. QUAD9 is seen to have the extra node at the centre compared to QUAD8, and for higher orders, more internal nodes are required. The polynomials are complete per order and are fairly competitive with the serendipity ones. However, a large percentage of the nodes in a mesh are the internal nodes, which is not always desired.

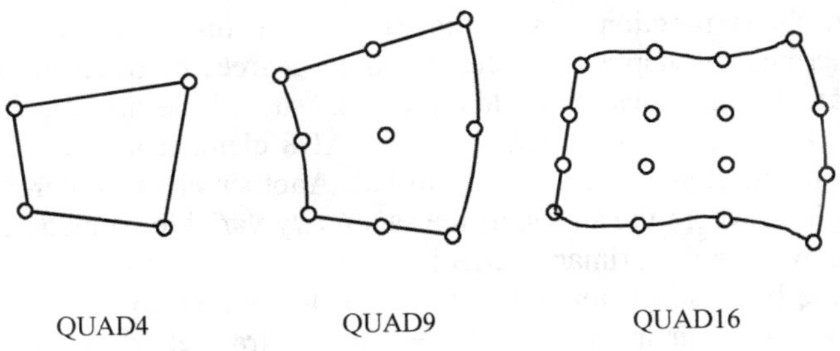

Figure 2.3 2D Lagrangian Element Family

2.2.3 Triangular Elements

Triangular shaped elements again follow an increasing order of shape functions (figure 2.4). TRI3 is the constant stress element which dates back to the start of finite element development, simple to implement but unfortunately a poor performer. These elements again use complete polynomials for each order. Triangles have the advantage of fitting into awkward shapes more easily than quadrilaterals, and so are more usual in some kinds of automatic mesh generation. They are also used in predominantly quadrilateral meshes for stepping up or down levels of refinement or around complicated geometric features such as holes, re-entrant corners, and crack tips.

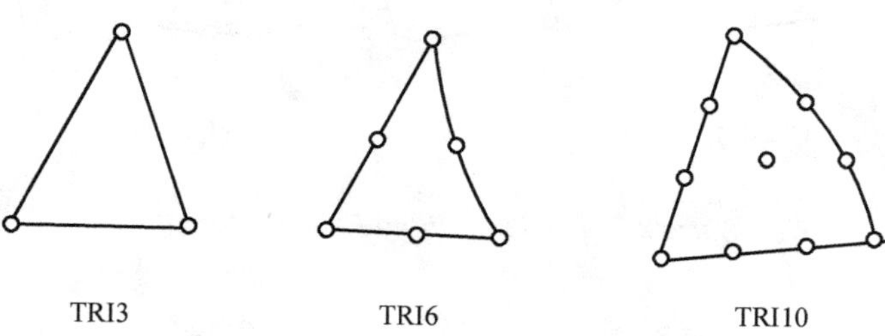

Figure 2.4 2D Triangular Element Family

2.2.4 Other Element Families

Other types of 2D elements have been developed in the quest for economic all-round accuracy. Instead of using additional nodes along the sides to raise the order of the shape functions, derivatives of the primary variables can be added as extra degrees of freedom at the corner nodes. Use is made of Hermitian polynomials. However, the derivative degrees of freedom can cause problems in boundary condition definitions. Sometimes, element performance can be enhanced by adding extra internal shape functions, as in the case of the **incompatible** variation of QUAD4, described in chapter 3. Here, the main degrees of freedom cease to be compatible along the sides, but although this should be illegal, in fact the performance becomes closer to that of the QUAD8 element in many cases. This breaks down if the element shape is distorted. Another element family includes stress equilibrium, where the stress, as the secondary variable, is made compatible along sides as well as the primary variable. Again, this is equivalent to increasing the order of the basic shape functions. Lastly, it is possible to introduce a higher order of strain behaviour in the so-called enhanced strain elements, a feature that has been successful when used in material and geometric non-linear analysis.

Elements can also be used in 1D, as line elements used typically for reinforcing and contact control. There are two nodes for linear displacement variation, three for quadratic, and so on, although the number of degrees of freedom per node can vary depending on the problem type. In 3D space, 3 would be required, whereas just one (temperature) is required for thermal calculations, and more than three (including rotations) for beams and shells. The element BAR2 is a typical 1D element with one degree of freedom per node, shown later in figure 3.1.

The 3D elements are a straightforward extension of 2D ideas, although many more element types can be devised. A selection is shown in figure 2.5, again using the consistent nomenclature convention. Those shown are all serendipity elements, the Lagrangian ones requiring extra mid-face and internal nodes as appropriate. The WEDGE15 element is a mixture of triangular and quadrilateral faces, has quadratic displacement variation, and is useful for coarse to fine mesh grading in the same manner as the 2D triangle TRI6.

The isoparametric elements are very versatile and can be used without the need for excessive numbers in practical applications, certainly for quadratic elements upwards. Distorted shapes can be tolerated within reason. In order to appreciate how these elements perform and hence how best to use them, some theoretical considerations have to be understood, as described below. For simplicity in the rest of this chapter, most of the mathematical arguments refer to 2D, although they are directly extendible to 3D.

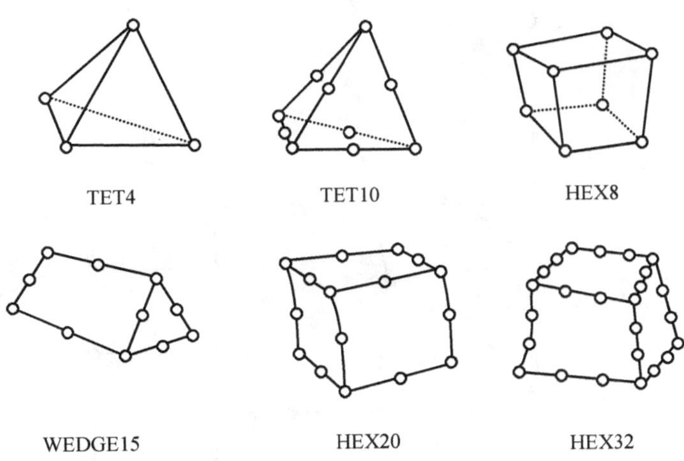

Figure 2.5 A Selection of 3D Element Types

2.3 Basic Element Behaviour

An understanding of basic element behaviour requires certain theoretical developments that are given in the following sections. Much of this is from standard theory, which is detailed in standard texts, e.g. [8] and [9].

A shape function is used to define any primary field variable in an element. This can be written as

$$\theta = \Sigma p_i \theta_i \qquad (2.1)$$

where the suffix i indicates values at the nodes and summation is over the nodes on this element; θ is the primary variable which can be any component of displacement, pressure, geometry, or temperature, and p_i are the actual shape functions.

The shape functions are polynomial representations of behaviour inside each element. This behaviour is written in terms of local element coordinates. Consider 2D behaviour: 3D is a straightforward extension. Although there are differences in detail in the theoretical derivations between triangles and quadrilaterals, the end results are equivalent, and it is more instructive to follow those of the latter. Consider, therefore, any quadrilateral, assumed to have n nodes

The basic theory is developed in a square of side length 2 units, which can be termed the **fundamental** shape. This space is the **theory space** and has dimensionless coordinates (ξ,η). The real, or **user space**, has the usual (x, y) cartesian coordinates. These spaces are shown in figure 2.6 for the QUAD4 and QUAD8 quadrilaterals.

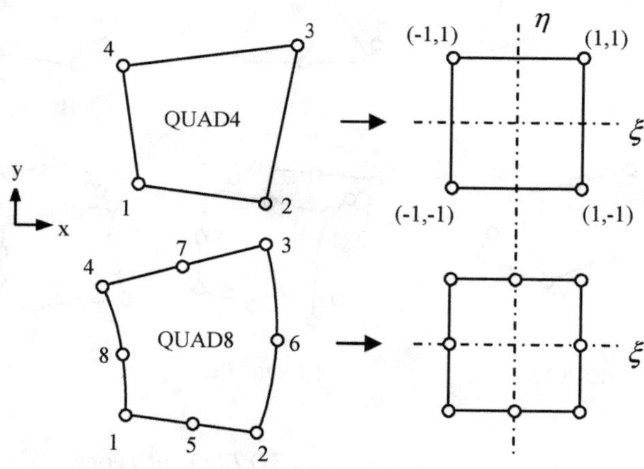

Figure 2.6 The Transformation Between Real (Left) and Theory (Right) Spaces

If the degrees of freedom at each node i are the displacement components (u_i, v_i), as used in stress and dynamics analysis, then the shape functions can express (u, v) at any other point in the element based on these nodal values as:

$$u = \Sigma p_i u_i \qquad v = \Sigma p_i v_i$$
$$u = p_1 u_1 + p_2 u_2 + ... + p_n u_n \tag{2.2}$$
$$v = p_1 v_1 + p_2 v_2 + ... + p_n v_n$$

where p_i are the shape functions and the summation is over the nodes on this element. There is a shape function for every node, so that p_1 refers to node 1, etc. A requirement of each of these shape functions is that they equal unity at their reference node and zero at every other node. Also, $\Sigma p_i = 1$ throughout the element.

The shape functions for the linear and quadratic displacement elements QUAD4 and QUAD8 from the serendipity family, and QUAD9 from the Lagrangian family, are given in Table 2.1. Node ordering is as shown in figure 2.6, with the centroidal node of QUAD9 being number 9. Also shown are the local coordinate values of each node. Thus, for QUAD4, node 1 has (ξ, η) coordinates (-1,-1), and

$$p_1 = \tfrac{1}{4}(1-\xi)(1-\eta) \tag{2.3}$$

Although the details are not too important to the finite element user, useful insight into element behaviour comes from writing the shape function in terms of its polynomial terms, as

$$p_i = a_1 + a_2\xi + a_3\eta + a_4\xi\eta \tag{2.4}$$

where the a_i are functions of the nodal geometry, for each of the four nodes, *i*, on the element. The number of terms in the polynomial equals the number of nodes. A more compact way of writing the polynomial dependence is

$$p_i = (1, \xi, \eta, \xi\eta) \tag{2.5}$$

It can be seen that this is linear due to the presence of ξ, η, and has bilinear dependence (a pseudo quadratic effect) due to the presence of $\xi\eta$. The **order** of this polynomial is therefore unity, which is the highest power over the coordinates in equation (2.5). Thus, QUAD4 is sometimes called a **bilinear** element, and its 3D equivalent HEX8 a **trilinear** element.

For the eight-noded quadratic displacement element QUAD8, there are eight terms in the shape function polynomial (Table 2.1), which is now of order 2:

$$p_i = (1, \xi, \eta, \xi\eta, \xi^2, \eta^2, \xi^2\eta, \xi\eta^2) \tag{2.6}$$

and, as above, each shape function equals unity at its reference node and zero at every other node. The shape functions now show linear, quadratic and weak cubic dependence (biquadratic).

Table 2.1 *Shape Functions of 4, 8 and 9 Noded Quadrilaterals*

Coords		shape function	*include only if node i exists on the element* i=5	i=6	i=7	i=8	i=9
-1	-1	$p_1 = \frac{1}{4}(1-\xi)(1-\eta)$	$-\frac{1}{2}p_5$			$-\frac{1}{2}p_8$	$\frac{1}{4}p_9$
1	-1	$p_2 = \frac{1}{4}(1+\xi)(1-\eta)$	$-\frac{1}{2}p_5$	$-\frac{1}{2}p_6$			$\frac{1}{4}p_9$
1	1	$p_3 = \frac{1}{4}(1+\xi)(1+\eta)$		$-\frac{1}{2}p_6$	$-\frac{1}{2}p_7$		$\frac{1}{4}p_9$
-1	1	$p_4 = \frac{1}{4}(1-\xi)(1+\eta)$			$-\frac{1}{2}p_7$	$-\frac{1}{2}p_8$	$\frac{1}{4}p_9$
0	-1	$p_5 = \frac{1}{2}(1-\xi^2)(1-\eta)$					$-\frac{1}{2}p_9$
1	0	$p_6 = \frac{1}{2}(1+\xi)(1-\eta^2)$					$-\frac{1}{2}p_9$
0	1	$p_7 = \frac{1}{2}(1-\xi^2)(1+\eta)$					$-\frac{1}{2}p_9$
-1	0	$p_8 = \frac{1}{2}(1-\xi)(1-\eta^2)$					$-\frac{1}{2}p_9$
0	0	$p_9 = (1-\xi^2)(1-\eta^2)$					

The presence of these bilinear and biquadratic terms in the four-noded and eight-noded quadrilaterals are responsible for some of the interesting properties they have, such as enhanced accuracy with reduced integration, as described later. Such terms are missing in the linear and quadratic triangles, which therefore perform less well. The shape function derivations for these latter elements are a little different to the above, but readily fit into the overall element developments.

The quadratic Lagrangian element QUAD9 contains the extra polynomial term $\xi^2\eta^2$ in the above equation for QUAD8. The performance is therefore a little better, although the enhanced accuracy gains are not so apparent.

Equivalent arguments apply in 3D and in other isoparametric types of element, in particular in some shell formulations. The main difference for 3D is that the theory space now has the three dimensionless coordinates (ξ, η, ς), although the orders of the respective polynomials for linear, quadratic, and so on will be the same.

The equations (2.2) show the main use of the shape functions, which is to calculate the value of any field variable at any point (ξ, η) within the element from its nodal values. In a structural analysis, after assembly of all the elements and performing a structural solution, the nodal displacement values will be evaluated in a manner that upholds the shape function variations. For ease of notation, the two degrees of freedom u and v at any point in the element are linked together as a vector $\{\delta\} = \{u, v\}$ in 2D, or $\{\delta\} = \{u, v, w\}$ in 3D, so that $\{\delta\}$ is general to any dimension.

Then, if the vector of corresponding nodal values of $\{u, v\}$ over the element is $\{\delta\}^e$, then the vector equivalent of equation (2.1) is:

$$\{\delta\} = [N]\{\delta\}^e \tag{2.7}$$

By differentiation, it is also possible to calculate the strains at the same point:

$$\{\varepsilon\} = [B]\{\delta\}^e \tag{2.8}$$

Here, $[B]$ is a matrix of shape function derivatives and $\{\varepsilon\}$ is the strain vector at this point. Thus, $\{\varepsilon\} = (\varepsilon_x, \varepsilon_y, \tau_{xy})$ and these three strain components with the three corresponding stress components are the only active ones in 2D plane stress and plane strain. Although the out-of-plane strain (plane stress) and out-of-plane stress (plane strain) are non-zero, respectively, their product is zero and so they do not contribute to the strain energy. For axisymmetric formulations, a fourth, hoop, stress component exists along with a hoop strain, which produces a non-zero contribution to the strain energy. The present discussion is restricted to plane stress and plane strain.

From the compatibility relationship of strain to displacement, $\varepsilon_x = \partial u / \partial x$, etc., since $u = \Sigma p_i u_i$, we need to express the strain components in terms of the shape functions to evaluate the terms of $[B]$. After some manipulation, which in detail is not necessary here, it turns out that $[B]$ can be written out in full as:

$$[B] = \begin{bmatrix} d_{11} & 0 & d_{12} & 0 & \vdots & 0 \\ 0 & d_{21} & 0 & d_{22} & \vdots & d_{2N} \\ d_{21} & d_{11} & d_{22} & d_{12} & \vdots & d_{1N} \end{bmatrix} \tag{2.9}$$

where the three rows correspond to the three components of strain, and the columns cover the N nodes. A typical term applying to node k is

$$\begin{aligned} d_{1k} &= \left[\left(y_i \frac{\partial p_i}{\partial \eta}\right)\frac{\partial p_k}{\partial \xi} - \left(y_i \frac{\partial p_i}{\partial \xi}\right)\frac{\partial p_k}{\partial \eta}\right] / \det J \\ d_{2k} &= \left[-\left(x_i \frac{\partial p_i}{\partial \eta}\right)\frac{\partial p_k}{\partial \xi} + \left(x_i \frac{\partial p_i}{\partial \xi}\right)\frac{\partial p_k}{\partial \eta}\right] / \det J \end{aligned} \tag{2.10}$$

An important factor here is the inverse det$J$ term. The scalar det$J$ is the Jacobian transformation, effectively a numerical representation, or scaling, between the theory space and the real space of the element, over which the integration is performed. It is described later in this chapter. Each of the above derivative terms of the shape functions p_i are polynomials of one order less than p_i.

The corresponding stress vector is $\{\sigma\}$ and using Hooke's law, the stresses can be related to the strains at the same point in the element by

$$\{\sigma\} = [D]\{\varepsilon\} \tag{2.11}$$

[*D*] is a matrix of combinations of the material constants *E* (Young's modulus) and ν (Poisson's ratio). If the element properties are isotropic, then [*D*] contains many zero terms. For anisotropy, [*D*] will be more populated which can affect certain aspects of element behaviour.

The most common finite element method is based on the principle of minimum potential energy, which means that when the structure is loaded and suitably fixed, the displacements and stresses are those which render the potential energy a minimum. Other equivalent principles lead to the same result, which requires the solution of the **stiffness equations**:

$$[K]\{u\} = \{f\} \tag{2.12}$$

Here, [*K*] is the global stiffness matrix, {*u*} is the total degree of freedom vector, and {*f*} is the corresponding vector of equivalent nodal loads.

Each element contributes its own stiffness matrix, $[K]^e$, added into the global stiffness matrix, plus any contributions to the load vector. Prescribed displacements are specified separately. The **element stiffness matrix** is given by:

$$[K]^e = \int_{-1}^{1}\int_{-1}^{1}[B]^T[D][B]\det J d\xi d\eta \tag{2.13}$$

in terms of the element's 2D theory space.

2.4 Numerical Integration

The integration in equation (2.13) cannot be performed analytically, because the equations are too complicated in all but the simplest types of elements. Hence, a numerical integration scheme called **Gaussian quadrature** is used. The scheme integrates a polynomial curve in 1D, which effectively calculates the area under the curve over the required interval. The **rule order** is the number of integration points, or **Gauss points**, needed in the interval to perform an accurate numerical integration. For each order, the Gauss points have to be situated at particular locations within the interval. For a rule with *n* such points, polynomials of up to and including order *2n-1* are integrated accurately. In numerical analysis terms, Gaussian quadrature is very efficient, requiring less points than other methods of integration. Figure 2.7 shows the concept. In the limit of a straight line, *n=1*, the integration calculates the area of a simple quadrilateral. Polynomials higher than *2n-1* cannot be integrated accurately, which in finite element terms results in instabilities such as zero energy modes, as described in chapter 4. If a given polynomial is integrated accurately by a rule of order *k*, it is also integrated accurately by all orders higher than *k*.

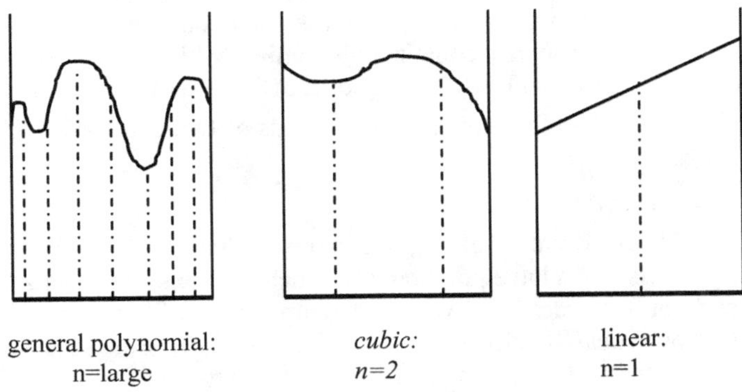

Figure 2.7 Numerical Integration using Various Orders of Polynomial

These integrations are evaluated over intervals which are 1D, lying in some particular spatial direction. Since finite elements cover 2D and 3D, the 1D numerical integration scheme is simply repeated with the intervals now in the 2 or 3 global directions, respectively. Figure 2.8 shows the arrangements of Gauss points in 2D quadrilaterals for rule orders 1 to 3, which are the most common ones used in practice. The locations of the points are indicated with respect to coordinates in the theory space. In 2D, the rules are written as (*n*x*n*), and in 3D as (*n*x*n*x*n*).

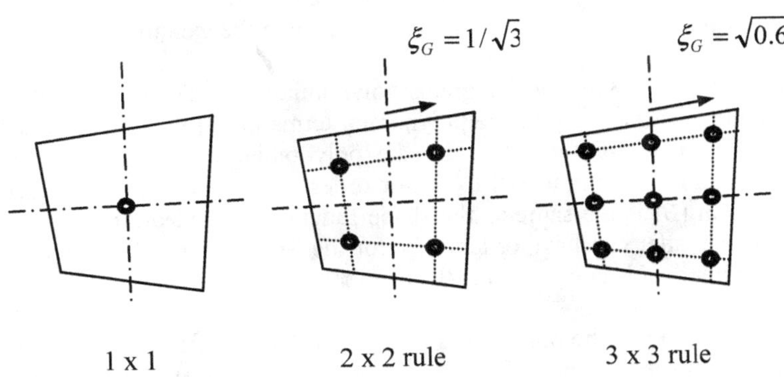

Figure 2.8 Gauss Point Locations for the Common Rules in Quadrilaterals

Although in 2D and 3D the polynomial to be integrated now contains multi-dimensional terms, the integration is still accurate provided the polynomial order is *2n-1* or less. Hence, the polynomial form of the stiffness matrix will indicate the required quadrature rule. Consider the expansion of the integrand of (2.13) at each Gauss point. Each term in the matrix $[D]$ is assumed constant (isotropic materials predominate), but those in $[B]$ contain polynomials which are derived from the shape functions, as in equations (2.10) above. All the terms in the triple matrix product $[B]^T[D][B]$ therefore contain polynomials of twice the order of the terms in $[B]$:

$$[B]^T[D][B] = (1, \xi, \eta, \xi^2, ...) \tag{2.14}$$

Because of the 2D or 3D nature of $[B]$, due to the products of derivatives in pairs in equations (2.10), its terms have polynomials which are the same order as the shape functions, and not one order less. An added complication of $[B]$ is that all its terms are multiplied by $1/\det J$, as also shown in equation (2.10). Since the inverse of a polynomial is an infinite polynomial, if $\det J$ is not constant, its inverse is also an infinite polynomial, and so is every term of $[B]$ and consequently the product terms of $[B]^T[D][B]$ in equation (2.14).

When $\det J$ is constant, i.e. when the element is relatively undistorted (see section 2.7), suppose the polynomial order of equation (2.14) is N. Then if a rule of order n is chosen, provided $N \le 2n-1$, the numerical integration is exact and we can term the rule a **complete integration**. It so happens that the rule that is one order less, i.e. n-1 in each direction, can result in much greater accuracy, although technically the numerical integration is no longer exact. Such a rule is known as **reduced integration**. If $\det J$ is not constant, when the element is now distorted, the polynomial of equation (2.14) is infinite and so in theory an infinite rule is required. However, in practice, the complete and reduced rules that apply when $\det J$ is constant are assumed to be applicable here as well. Tests can be made to show these rules are sufficiently accurate, as shown in the example in section 2.6.

As a 2D example, a 3x3 rule integrates polynomials of order up to 5 accurately. This is obtained if the order of the polynomial terms in $[B]$ is 2 or less and $\det J$ is constant. Similarly, for the 2x2 rule, the polynomial order of 3 is accurately integrated requiring terms in $[B]$ to be for order 1, i.e. linear. Since the order of polynomials in $[B]$ is the same as the shape functions for quadratic elements, this shows that the latter can have orders of 2 for the 3x3 rule, and 1 for the 2x2 rule, again assuming $\det J$ is constant over the element.

Note that in the above, the polynomial of the terms arising from the matrix product $[B]^T[D][B]$ has orders 4 and 1 for the 3x3 and 2x2 rules respectively, each being one less than the $2n$-1 value. Hence the products are in each case one order less than that which could be integrated accurately, and which reflects a slight loss in integration efficiency.

This establishes relations between element shape function orders and a suitable complete integration rule. For quadratic 2D and 3D elements whose shape functions are of order 2, a complete rule of order 3 applies along with a reduced rule of order 2. These rules may be too low if anisotropic material behaviour exists, since then the [D] matrix may no longer be constant over the element, when extra geometric terms would be added into the triple matrix product of equation (2.14).

It is also apparent that for an analytical evaluation of the stiffness matrix to be possible, det$J$ must be constant, although this is rarely used these days.

The Gaussian quadrature integration rules suitable for different members of the isoparametric element families are shown in Table 2.2. Note that, for triangles and tetrahedra, the numerical integration rules given are as for the corresponding quadrilaterals and hexahedra. However, for these element families the complete rule is often replaced by a special triangle rule (e.g. 3 points at the midside nodes for TRI6). Optimal points do not exist for these families, as discussed in section 3.4, so the way such points are interpreted will depend on individual software.

Table 2.2 *Basic Element Properties*

		Number of:		El Order	Num Intgn Rules:		Stress Var'n:	
	El Type	*nodes*	*d of f*		*complete*	*optimal points*	*complete*	*pseudo*
1D	BAR2	2	2	linear	1	1	const	-
2D	TRI3	3	6	linear	1	1	const	-
2D	TRI6	6	12	quadr	3x3	2x2	linear	-
2D	TRI10	10	20	cubic	4x4	3x3	quadr	-
2D	QUAD4	4	8	linear	2x2	1x1	const	linear
2D	QUAD8	8	16	quadr	3x3	2x2	linear	quadr
2D	QUAD9	9	18	quadr	3x3	2x2	linear	quadr
2D	QUAD12	24	24	cubic	4x4	3x3	quadr	cubic
3D	TET4	4	12	linear	1x1x1	1x1x1	const	-
3D	TET10	10	30	quadr	3x3x3	2x2x2	linear	-
3D	WEDGE15	15	45	quadr	3x3x3	2x2x2	linear	quadr
3D	HEX8	8	24	linear	2x2x2	1x1x1	const	linear
3D	HEX20	20	60	quadr	3x3x3	2x2x2	linear	quadr
3D	HEX27	27	81	quadr	3x3x3	2x2x2	quadr	cubic
3D	HEX32	32	96	cubic	4x4x4	3x3x3	quadr	cubic

The optimal points are also the reduced rule, although the latter may give mechanisms for some of the element types

2.5 Reduced Integration

Having established what the complete rule is for a given element type, of given shape function order and element shape, the reduced rule is one order less. The convergence of results to the exact solution with mesh refinement is assured for any integration rule, n, if that rule integrates the volume of the element exactly.

This is because, as a mesh is refined and each element represents a smaller volume, the strain energy density, ρ, tends to a constant value in each element. But this is also the quantity that is integrated over each element to give the element stiffness matrix in 3D as:

$$\rho^e = \int_e \rho dV = \int_{-1}^{1} \int_{-1}^{+1} \int_{-1}^{+1} \rho \det J d\xi d\eta d\varsigma \tag{2.15}$$

As ρ becomes constant, it can be written outside the integral, which then becomes simply the integration of the element volume (q.v. equation 2.18). Thus, the integration rule only has to integrate the volume properly.

Hence, if the order of the det$J$ polynomial is N and the rule is n, then n has to be such that $2n-1 \geq N$. But det$J$ is a relatively low-ordered polynomial, the more so the less distorted the element shape, so a relatively low rule would suffice here. In fact, another feature of ultimate mesh refinement is that curved quadrilateral elements tend to parallelogram shapes, and curved triangles tend to straight-sided triangles, and det$J$ is constant in both cases. So this integration becomes that of a constant, and therefore a single Gauss point rule performs the integration exactly.

If the rule is too low, so that $2n-1 < N$, the stiffness matrix is numerically indefinite and spurious results, often in the form of excessive displacements, occur. These are generally referred to as **instabilities** or **mechanisms**, and the deformed element shapes are termed **zero energy modes** or **hourglass modes** (section 4.8). The reduced rule is normally higher than this rule, and should avoid such mechanisms and give more rapid convergence than the complete rule, since it produces a less stiff stiffness matrix. The reduced rule Gauss points are situated at the **optimal stress points**. Because the secondary variables (stresses and strains) are one order less than the shape functions, their accuracy varies over the element. They are at their most accurate at these optimal stress points, also known as **Barlow points** [12], whatever the degree of distortion. Since the stiffness matrix is evaluated from the product of strain terms, it follows that its integration using these optimal points should produce a more accurate stiffness matrix than if integrated over other Gauss points, including those of the complete rule. This conclusion holds as long as the reduced rule does not produce adverse effects such as mechanisms.

Many tests, particularly of bending in shells and solids, have shown much better convergence rates, i.e. greater accuracy for a given mesh, with the reduced rule compared to the complete rule. However, reduced integration is only really effective for quadrilaterals in 2D and hexahedra in 3D, and not for triangles in 2D and tetrahedra and wedges in 3D, since those elements do not have optimal points. For convenience for such elements, the reduced rule Gauss points are often used as locations to calculate the stresses, where accuracy should still be better than at the nodes. This maintains consistency with the quadrilaterals and hexahedra.

Selective integration schemes have also been researched and provide useful alternatives, but the above complete and reduced rules are the most common ones available in commercial software for computing element stiffness matrices. The above arguments are mainly concerned with quadratic displacement elements in either 2D or 3D, but they also apply to higher ordered elements, although the reduced rule gains are not so dramatic.

2.6 Example of Numerical Integration Aspects

As an example of some of these numerical integration issues, consider a square 2D shape comprising 4 quadrilateral elements of varying types (figure 2.9). Numerical integration rules between 1 and 5 are considered. Minimum fixings are applied to the left-hand side, to prevent rigid body motion, and a parabolic shear (or linear moment) load is applied upwards along the right-hand edge. This produces quadratic stress and cubic displacement variations throughout the square, and so is a severe test for such coarse meshes, the more so for low order elements. Node A is fixed in both directions and node B is restrained from horizontal displacement. The nodes in between are not fixed but have horizontal loads applied which should prevent horizontal displacement. Two versions of the mesh are considered, one with a regular pattern of square elements, the chain-dashed lines of figure 2.9, and the other a distorted mesh, shown dashed. The central node has been shifted to produce a noticeable distortion of amount often used in practice.

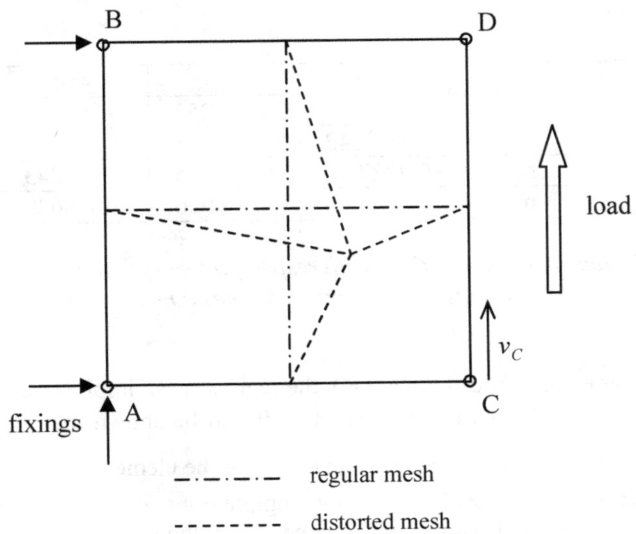

Figure 2.9 Mesh Used for Numerical Integration Tests

The 4 types of quadrilateral QUAD4, QUAD8, QUAD9 and QUAD12 are all used in both meshes, with all side nodes positioned in their correct locations. The maximum displacement is vertical and occurs at the right-hand end corners C and D (both the same), so that results for v_C only are shown in Table 2.3 for all element types and the two mesh variants. Most results are very accurate, the exceptions being when the integration rules are too low for the particular element. The term *mechanism* indicates excessive displacements, many orders of magnitude too high. The hourglass modes that are highlighted in the QUAD4 and QUAD9 reduced rules give these spurious mechanisms, which are of the same order of magnitude as the correct displacement modes, although the stresses at the optimal point are accurate.

Table 2.3 *Maximum Displacements in the Four Element Mesh under Parabolic Shear Loading*

(a) regular mesh

Element	1x1	2x2	3x3	4x4	5x5
QUAD4	1.160000*	0.917655	0.917655	0.917655	0.917655
QUAD8	m	0.998764	0.998363	0.998363	0.998363
QUAD9	m	0.994000	0.995358	0.995358	0.995358
QUAD12	m	m	1.000000	1.000000	1.000000

(b) distorted mesh

Element	1x1	2x2	3x3	4x4	5x5
QUAD4	m	0.891085	0.890251	0.890249	0.890249
QUAD8	m	1.012842	1.002708	1.002669	1.002669
QUAD9	m	0.922748*	0.994265	0.994243	0.994243
QUAD12	m	m	1.008976	1.008639	1.008636

All displacements normalised with respect to theory; an asterisk shows an hourglass distortion; m denotes a mechanism

For the linear element mesh (QUAD4), the regular mesh implies a constant det*J*, so that a 1x1 rule is the lowest reduced rule. It can be shown that the terms d_{ij} in equation (2.10) are linear in (ξ,η) and therefore the element stiffness matrix $[K]^e$ is quadratic, implying a 2x2 rule for the complete integration. Higher rules would all give the same result. However, in plane stress, the rank argument (described in section 4.8) fails for the 1x1 rule and the stiffness matrix is indefinite. This is apparent in the results, where hourglass distortions appear in the regular mesh, and mechanisms in the distorted mesh. In this latter, det*J* is now linear but still a 2x2 rule suffices for complete integration. The stiffness matrix has no exact integration, but as shown in Table 2.3, little difference occurs with the higher rules, so that the rule of thumb, that the complete integration should be as in the constant det$J$ case, is supported. Thus, for QUAD4, the 2x2 rule is complete but the reduced rule 1x1 does not work.

For the quadratic elements (QUAD8), with the square mesh, the terms d_{ij} of equation (2.10) are now quadratic in (ξ,η), so that $[K]^e$ is now quartic and a complete rule is 3x3. This is also assumed for the distorted mesh, as before. Table 2.3 shows that the regular mesh is indeed exactly integrated by 3x3 and higher rules. For the distorted mesh, the 3x3 rule gives almost the same result as the 4x4 rule, but does not, as QUAD4, give an exact integration. The 1x1 rule is totally inadequate but the 2x2 rule is a very accurate reduced rule for QUAD8.

The Lagrangian element QUAD9 shows an hourglass mode for the 2x2 reduced rule, otherwise it behaves like the serendipity QUAD8 element.

The cubic displacement QUAD12 element can represent the quadratic stress variations exactly, hence the exact results in Table 2.3(a). In this case, however, the 2x2 rule is too low and produces mechanisms. The 3x3 rule is a reduced rule and 4x4 the full integration rule, as shown in Table 2.2.

For all the element types, the numerical integration in the regular mesh is seen to be exact in the sense that the complete integration and all higher rules give identical results, so that the stiffness terms are all identical. In the distorted mesh, this is not so apparent, as discussed above, although the results using the 5x5 rule often equal those using 4x4.

2.7 Jacobian Transformations

A fundamental factor connecting real and theory spaces in each element is the Jacobian, a matrix given at any point in the element in 2D as

$$[J]=\begin{bmatrix}\dfrac{\partial x}{\partial \xi} & \dfrac{\partial y}{\partial \xi}\\ \dfrac{\partial x}{\partial \eta} & \dfrac{\partial y}{\partial \eta}\end{bmatrix} \tag{2.16}$$

The 3D version has the third row and column with the appropriate z and ς terms. Its 2D determinant is given by:

$$\det J=\frac{\partial x}{\partial \xi}\frac{\partial y}{\partial \eta}-\frac{\partial x}{\partial \eta}\frac{\partial y}{\partial \xi} \tag{2.17}$$

The Jacobian transformation is a matrix of each of the spacial derivative terms in equation (2.16), but the determinant is of more significance to the user. This determinant can be seen as an indication of how much each point stretches from the double unit space to the real space (known as a **mapping**). In a two noded 1D line element, det*J* would be simply the real space length divided by 2 (because the length in the theoretical space is two units). Because the 1D element is simple enough, det*J* is constant over all points in the element.

The area (in 2D) or volume (in 3D) of any part of space shows the significance of det*J* as a point-wise variable so that, in 2D for instance,

$$V = \int_V dxdy = \int_{-1}^{1}\int_{-1}^{1} \det J d\xi d\eta \tag{2.18}$$

Based on the shape functions, it is possible to write det*J* as a polynomial in terms of (ξ,η). This is because at any point (*x*,*y*) in the real space, the values of *x* and *y* can be written by shape functions in terms of (ξ,η). From there, the determinant can be worked out.

Regular shaped elements, such as rectangles, parallelopipeds, and straight-sided triangles, all have the same value of det*J* over the element. This means the polynomial for det*J* contains zero coefficients of ξ and η and higher terms. As the shape becomes less and less regular, i.e. more terms in the shape functions defining the geometry are non-zero, the order of the det*J* polynomial increases. The most general such form for the QUAD4 element is:

$$\det J = f(1,\xi,\eta) \tag{2.19}$$

and for QUAD8 is:

$$\det J = f(1,\xi,\eta,...,\xi^3\eta,\xi^2\eta^2,\xi\eta^3) \tag{2.20}$$

When det*J* is no longer constant, the stiffness matrix polynomial becomes of infinite order, and a strictly accurate numerical integration is not available. This can be seen from equation (2.14), whose right-hand side contains $(1/\det J)^2$, and equation (2.13), with detJ in the integrand. The product which gives $[K]^e$ is seen therefore to contain 1/detJ in the integrand, which is an infinite polynomial. We can term any element whose shape is such that det*J* is not constant over the element as **distorted**. Only undistorted elements can be accurately evaluated numerically.

If singularities exist in the mapping, det*J* becomes zero, corresponding to certain types of gross distortion and discussed in section 5.9. Large numerical errors could result. Another requirement of the mapping is that it should not be both positive and negative in the same element. If negative throughout the element, the mapping sign can be reversed by software (e.g. by reversing the topology order).

Over an element, if det$J$ is not constant but a polynomial, then a useful check on the shape is to compare the maximum and minimum values of det$J$, or their ratio. Hence, checking det$J$ is a valuable element shape check.

Some of these points are expanded in chapter 5.

2.8 Geometric Representation of det*J*

A geometric representation of det$J$ is available from its vector definition. Define the vector $\bar{r} = (x, y, z)$ and its vector derivatives $r_\xi = \dfrac{\partial \bar{r}}{\partial \xi}$, etc. Then det$J$ is a scalar resulting from the scalar triple product in 3D of r_ξ, r_η and r_ς. For 2D quadrilateral and 3D hexahedral (brick) elements, at any point in the element these vectors lie in the local space directions ξ, η, ς, respectively, which are in the local tangent directions ξ=constant, η=constant, ς=constant in the real space. Figure 2.10 shows this in a 2D case. This scalar triple product is

$$\det J = r_\xi . r_\eta x r_\varsigma \tag{2.21}$$

For 2D, r_ς is the out of plane unit vector. The interesting point about this definition is that several shape measures are involved in this definition (2.21). The relative magnitudes of the vectors r_ξ, r_η and r_ς define the **aspect ratio** at the same point, which we can term the **tangent vector aspect ratio**.

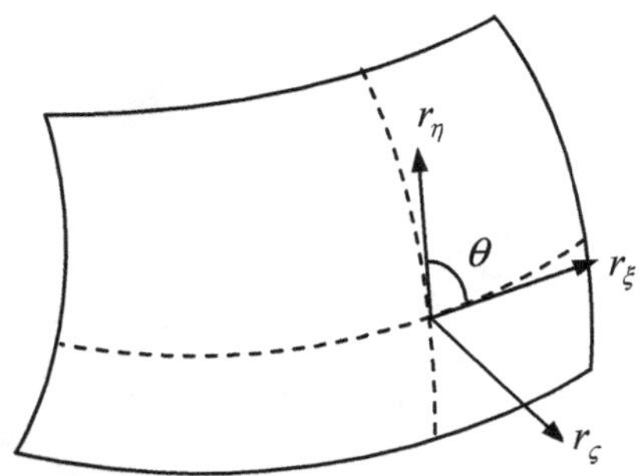

Figure 2.10 Definition of Tangent Vectors and Skew Angle per Reference Point

The largest vector magnitude divided by the smallest gives an indication of the aspect ratio at this point. The angle, or angles in 3D, between these vectors give the local **skew angle** or **skew angles.**

These aspect ratios and angles are all useful measures for describing the degree of distortion in each individual element.

2.9 Summary Comments

In this chapter, the main element families likely to be used in commercial software have been described. Aspects of element design and the basic theory pertaining to the more important types have been given in sufficient detail for an understanding of some important behavioural properties. For the isoparametric element families, these include how numerical integration works, with Gaussian quadrature rules that have to be of sufficient order to integrate the polynomial forms of the stiffness matrices. Both complete and reduced integration have been described, the latter being advantageous for quadrilateral and hexahedral elements in many situations. An example has highlighted some important numerical integration aspects. Since elements are required with different shapes in practical meshes, they have to be distorted from the simple theoretical shapes, and this distortion is monitored by Jacobian transformations. These are discussed both theoretically and geometrically, to give the user some understanding of how distortions may be measured. Their significance, and how errors can be introduced when excessive distortions are employed, is developed further in chapter 5.

3. Comparison of Main Element Types

3.1 Introduction

Having described the various properties of the different types of elements, it is now expedient to make comparisons of the main types in both 2D and 3D in the context of elastic stress analysis. Only the isoparametric element families are considered.

It has always been a common exercise for researchers and new users alike to investigate the relative performance of the different element types within each family by convergence tests. Such tests have shown that, for a given test case, convergence of both displacements and stresses occurs with increasing numbers of degrees of freedom for any element type. Also, the higher the order of element in terms of the shape functions it uses, the better its performance. Quadrilaterals (2D) and hexahedra (3D) are usually better than triangles (2D) and tetrahedra (3D). Many such experiments have been recorded in the literature, and some have been formalised in NAFEMS publications. The following sections discuss some of the reasons behind these performances.

3.2 Stresses

The variation of stresses within an element is an important consideration, since they are often the result required. A suitable distribution of some stress component across a part of the structure will therefore give a guide as to how many elements are needed there. The higher the order of element, the more stress reference points per element and therefore the fewer elements required to achieve this. Stresses give more information about performance than the displacement components from which they are derived, and so their study below is of more use.

We consider 1D, 2D and 3D families of isoparametric displacement elements. Since the displacements are given by assumed shape functions, equations (2.2), so are the stresses, which are geometric derivatives of displacements. In a 1D case, every component of stress can be written as a polynomial of the form:

$$\sigma = a + b\xi + c\xi^2 + d\xi^3 + \ldots \qquad (3.1)$$

Each term on the right hand side is a piece of information on stress variation over the element that comes, by virtual work considerations, from the available displacement components over the element. As an example, a simple 1D bar element BAR2 in a 1D space (figure 3.1) has a node at each end, with one degree of freedom per node and one component of rigid body motion along the line of the element. Thus, the element has two degrees of freedom, or one free degree of freedom when the rigid body mode is removed, available for deformation. Hence, there is a total of one stress component that can be evaluated in this element. This is clearly the constant component in the lengthwise direction of the bar.

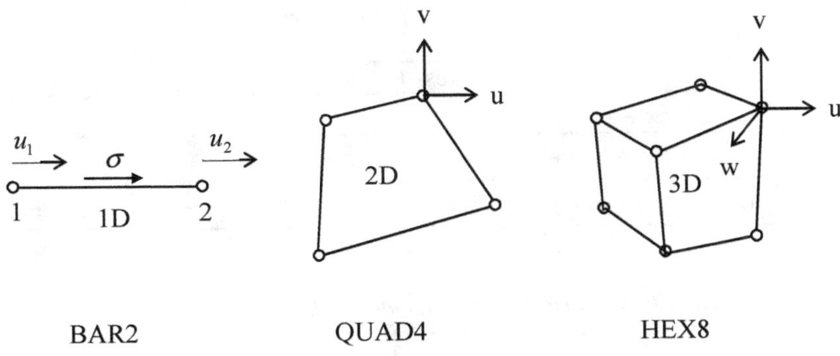

Figure 3.1 Basic Elements in each Spacial Dimension

In 2D and 3D, multi-directional components of geometry exist so the additional directional components of displacements and stresses have to be added into these calculations. Rules generalising the 1D case can be stated as:

- the number of available stress components per element = the number of *free* degrees of freedom (i.e. the number of element degrees of freedom less rigid body modes),
- the number of stress **sets** = integer part of stress components per element divided by the number of stress components per reference point in the element.

A **stress set** has been defined [13] as the number of terms in the polynomial representing each of the complete orders of variation over the element, i.e. constant stress, linear, quadratic, and so on. Table 3.1(a) shows which terms are involved in 1D, 2D (plane stress and plane strain) and 3D. Thus, for example, a 2D linear stress variation requires 3 terms in the polynomial and so has 3 sets. The same variation in 3D has 4 sets.

As an example, consider the 4-noded 2D quadrilateral QUAD4 (figure 3.1). There are two degrees of freedom per node and three rigid body modes in 2D. This gives a total of 4 x 2 – 3 = 5 free degrees of freedom, and therefore, stresses, over the element. But there are three components of stress per point, so there are int(5/3), or 1 complete stress set. This, by reference to Table 3.1(b), shows that constant stress is covered plus an incomplete (2/3) linear stress variation. Thus, QUAD4 permits pseudo-linear stress behaviour, which is better than constant but has incomplete linear behaviour. This incomplete part enhances the performance of this element type, as discussed below, and is the main reason why it performs so much better than the 3-noded triangle, which only permits constant stress.

Table 3.1(a) *Stress Polynomials in Different Dimensions*

Space	Stress Polynomial
1D	$\sigma = a + bx + cx^2 + dx^3 + \ldots$
2D	$\sigma = a + b_1 x + b_2 y + c_1 x^2 + c_2 xy + c_3 y^2 + \ldots$
3D	$\sigma = a + b_1 x + b_2 y + b_3 z + c_1 x^2 + c_2 xy + c_3 y^2 + c_4 yz + c_5 z^2 + c_6 zx + \ldots$

Table 3.1(b) *Stress Variations and Numbers of Stress Sets in Different Dimensions*

Space	Stress Variation	No. Stress Sets Required
1D	Constant	1
1D	Linear	2
1D	Quadratic	3
2D	Constant	1
2D	Linear	3
2D	Quadratic	6
3D	Constant	1
3D	Linear	4
3D	Quadratic	10

The above calculation can be extended to other key element types to give in each case the number of available stress sets. Table 3.2(a) gives basic dimensional properties and Table 3.2(b) shows the number of available stress sets for a range of element types along with the corresponding orders of stress variation. It is seen that the quadrilaterals and hexahedral elements all have incomplete contributions of stress polynomials, which are the same order as the displacement variations.

Table 3.2(a) *Basic Dimensional Properties*

Space	d of f per Node	Stress Comps per Ref Point	Rigid Body Modes
1D	1	1	1
2D	2	3	3
3D	3	6	6

Table 3.2(b) *Basic Element Properties*

	El Type	Number of: *nodes*	*free d of f*	No. Stress Sets: *complete*	*fraction*	Stress Variation: *complete*	*pseudo*
1D	BAR2	2	1	1	0	const	-
2D	TRI3	3	3	1	0	const	-
2D	TRI6	6	9	3	0	linear	-
2D	TRI10	10	17	5	2/3	quadr	-
2D	QUAD4	4	5	1	2/3	const	linear
2D	QUAD8	8	13	4	1/4	linear	quadr
2D	QUAD9	9	15	5	0	linear	quadr
2D	QUAD12	24	21	7	0	quadr	cubic
3D	TET4	4	6	1	0	const	-
3D	TET10	10	24	4	0	linear	-
3D	WEDGE15	15	39	6	1/2	linear	quadr
3D	HEX8	8	18	3	0	const	linear
3D	HEX20	20	54	9	0	linear	quadr
3D	HEX27	27	75	12	1/2	quadr	cubic
3D	HEX32	32	90	15	0	quadr	cubic

3.3 Constant Stress versus Linear Stress Elements

Stress variations across structures can vary in arbitrary ways, but over small enough sections can be considered to be more or less linear, in the sense that the error between linear, quadratic and higher variations would always be small over each element. Here, linear stress elements will behave well whereas constant stress elements will not. A good example of this is along a beam in bending, where the predominant stress component variation is linear. Suppose this stress is given by $\sigma = kx$ over a section of the beam where x varies from -1 to $+1$ (specific units are omitted from this example). Modelling with just one linear stress element across this section will give the correct answer. However, the response of constant stress elements depends on how many are spread across the section. Here they are assumed equal in size. Figure 3.2 shows schematically what and where the stresses in these elements are when $k = 1$. The maximum stress obtained from the constant stress elements (from the right-most element) is shown in Table 3.3 for increasing numbers of elements, along with the error in this stress. Also shown is the error in strain energy, which is easily calculated knowing that it is proportional to the integral of σ^2 across the section. In both cases, the theoretical values are reached slowly with increasing numbers of elements. In fact, if h is the element size in the x direction, the error in strain energy is proportional to h^2 and the error in the maximum stress is proportional to *h*. Hence, stress is more sensitive to error than the strain energy. It is seen that 3 elements produce a stress error of 33% and even 20 elements produce an error as much as 5%.

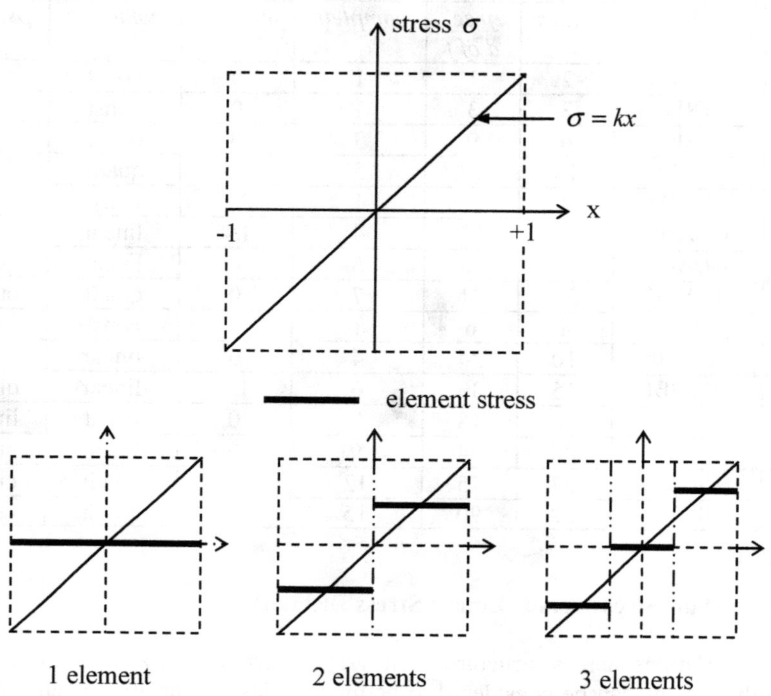

Figure 3.2 Stress Distribution for Constant Stress Elements in a Linear Stress Gradient

Table 3.3 *Convergence of Constant Stress Elements in a Linear Stress Gradient*

No. els	Max Stress	Error in: *max stress*	*strain energy*
1	0	1	1
2	1/2	1/2	1/4
3	2/3	1/3	1/9
4	3/4	1/4	1/16
n	$(n\text{-}1)/n$	$1/n$	$1/n^2$
theory	1	0	0

In this 1D example, the difference in performance between constant and linear stress (linear and quadratic displacement) elements is quite extreme. In 2D and 3D, the correspondingly poor results would only be reproduced by those elements that admit constant stress, TRI3 and TET4. The pseudo-linear stress availability in QUAD4 and HEX8 would render improved results. For higher stress variations across such a section, differences between linear and quadratic stress elements would be much smaller, but again constant stress elements would perform badly. These concepts equally apply in 1D, 2D and 3D.

In general, real world problems have complicated stress variations that are of higher order than linear across sections. Hence, linear and higher order stress elements would be required in sufficient numbers. Even so, much larger numbers of constant stress elements would be needed. The above example is 1D; when going from 1D to 2D, and 2D to 3D, the number of constant stress elements required will increase dramatically.

As a general rule, the gains of increasing the order of polynomials of stresses decrease significantly when going from constant to linear, linear to quadratic, quadratic to cubic, etc. From experience, the optimal order for stresses appears to be linear.

3.4 Triangles and Tetrahedra versus Quadrilaterals and Hexahedra

Considering 2D use, triangular elements have an advantage over quadrilaterals in filling space in self-adaptive re-meshing algorithms. However, their performance is not quite so good. This can be shown by considering the stress variations that were derived above. Both the 6-noded triangle TRI6 and 8-noded quadrilateral QUAD8 permit quadratic displacement and linear stress variations. TRI6 has 9 free degrees of freedom, so 9 stress components that, therefore, permit 3 stress sets and an exact linear stress variation. QUAD8 has 13 free degrees of freedom, so 13 stress components which therefore permit 4 stress sets, a linear stress variation, plus an extra quadratic component. QUAD9 has 15 free degrees of freedom, permitting 5 stress sets, also a linear stress variation plus two extra quadratic components although the latter order is again not complete.

These extra components cannot permit a complete quadratic stress behaviour, but they do provide sufficient information to detect where the "real" quadratic stress variations have their zeros. These are at the optimal stress points, as mentioned in section 2.5. Thus, if the stresses are evaluated at these points (often known as **Gauss point stresses,** the 2 x 2 rule being implied here), the stresses will be as accurate as if they had complete quadratic variation. For QUAD8 and QUAD9, these optimal points coincide with the 2 x 2 Gaussian quadrature locations. The triangle does not have this extra information and therefore no optimal points exist having this enhanced accuracy. Hence, a mesh of quadrilaterals will perform better than when each quadrilateral is replaced by two triangles (all with midside nodes). Another closely-related advantage is the use of quadrilaterals with reduced integration, where these optimal points coincide with the integration points. Note, however, that the performance of a predominantly quadrilateral element mesh will not be seriously impaired when a relatively smaller number of triangles are included, particularly when the triangles are use for stepping up the level of mesh refinement.

The elevated accuracy of the 2 x 2 Gauss point stresses in the quadrilateral elements is, unfortunately, not guaranteed elsewhere in those elements, and a typical variation with several elements along a stress gradient is shown qualitatively in figure 3.3. The best element fit curve is obtained by joining up smoothly the Gauss point stresses: this will not be the exact solution but is the best that can be obtained from the current degree of mesh refinement. For one element, the figure shows the linear extrapolation from the optimal points, the error e_1 this has at the element ends compared to the best element fit, and the error e_2 that the calculated element stresses have at the element ends (i.e. nodal stress results). Nodal averaged stresses will often be a good approximation, although extrapolated Gauss point stresses will give a more accurate average.

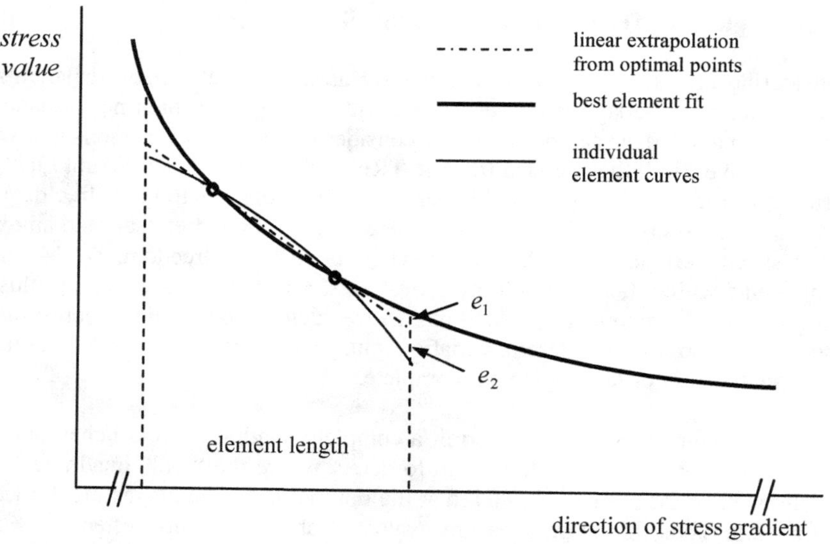

Figure 3.3 Schematic of Element Stress Errors at Element Ends

The same arguments apply for the quadratic hexahedra HEX20 and HEX27, and 10-noded tetrahedron, TET10, elements, the two former having optimal points at the 2 x 2 x 2 Gaussian quadrature locations.

For higher element types with higher stress variations, similar arguments exist but the gain in accuracy is not so significant.

3.5 Incompatible Elements

3.5.1 Poor Performance of Linear Displacement Elements

Both the 2D and 3D elements, QUAD4 and HEX8, have certain benefits since they are of simple form and only have corner nodes. These are apparent in some non-linear formulations, such as contact analysis, where nodes along sides complicate the physical behaviour. However, general performance is not good compared with the higher order elements and so very fine meshes are often indicated.

The elements are particularly poor performers in states of pure bending. In the case of a bending moment applied to a rectangular QUAD4, as shown in figure 3.4(a), the actual deformed shape is that of figure 3.4(b) whereas it should be as in figure 3.4(c). The latter reflects pure bending and no shear strains are present, unlike the attained state of QUAD4 in figure 3.4(b), which contains non-zero shear strains. Such strains are called **parasitic shear** and grow in magnitude as the element side aspect ratio (width to height) increases. This creates **mesh locking**, a term used to indicate that a correct solution cannot be obtained because the stiffness is artificially high for some reason. This effect should reduce with mesh refinement for all isoparametric elements, but can remain high in some circumstances when using plate and shell elements. These topics are discussed in detail in [8] and [11].

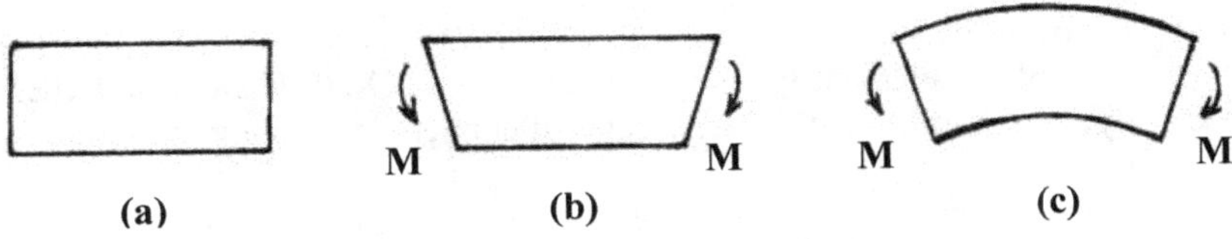

Figure 3.4 The Poor Performance of Linear Elements

The same adverse effect is observed in quadratic displacement elements, although to a lesser degree. The use of reduced integration (section 2.5) is very effective in removing these parasitic shears for all orders of quadrilateral and hexahedral elements, although not so effective for linear displacement elements. Hence, alternative solutions have been sought, particularly as described below.

3.5.2 Improvements Using Incompatible Formulations

Removing parasitic shears has inspired much research into the linear elements, of which the most successful is the incorporation of the so-called **bubble function**. This form is available in certain commercial software. This function is additional to the standard shape functions of equation (2.4), and adds to them terms proportional to $(1-\xi^2)$ and $(1-\eta^2)$, plus $(1-\varsigma^2)$ in 3D. This renders the element incompatible, or non-conforming, since elements sharing a common side may have different values of these additional terms along that side.

The result is to enhance the accuracy of these elements to almost that of the quadratic displacement elements. No extra nodes or degrees of freedom are involved since all the extra computations are conducted within each element's formulation. The disadvantage of the bubble function approach is that this accuracy is only available when det*J* is constant within every element in the mesh. Per element, the formulation requires the Jacobian transformation to be constant so as to pass the patch test (chapter 6), so that when it is not constant, the Jacobian value at the centroid is used. The overall accuracy is impaired even if only a number of the total elements have a varying Jacobian. The more an element distorts (the more det*J* varies over it), the less the accuracy, although the element should remain more accurate than with the basic QUAD4 formulation. Some software actually removes the bubble functions and only uses the basic QUAD4 or HEX8 formulations in those elements when det*J* is found not to be constant.

3.5.3 Example of Incompatible Elements

An example of incompatible elements is illustrated in a 3D beam, figure 3.5. This is loaded by an edge shear of 1000 units along one of the top edges, and all the bottom face is fixed encastré. This ensures a complicated stress pattern in that region, beyond simple beam theory and a severe test of the two elements used. The incompatible element incorporating the bubble function is herewith labelled HEX8I. The beam was modelled using two elements, as shown in figure 3.5, and run with each of the element types HEX8, HEX8I, HEX20, HEX27 and HEX32, respectively. This implies that different rates of convergence should be achieved.

The results are shown in Table 3.4, as displacements at point A on the loaded edge, and the nodal stress at point B in the base. Both components are maximum values over the beam. The fixed base and short height ensures the stresses are larger and displacements lower than those due to simple beam theory, but, for comparison sake, theoretical results from Roark [14] are included in the Table.

In order to test distortion effects (varying det*J*) of incompatible elements, four nodes are moved up and down the edges as indicated in figure 3.5. Using this, three extra runs are included, one using the basic HEX8 elements and the other two using HEX8I elements. The first of these allows det*J* to vary freely over the element, which should impair accuracy, the second enforcing constant det$J$ by using centroidal values.

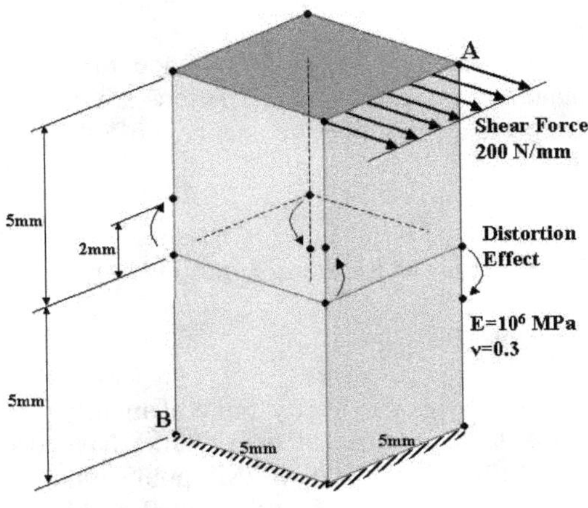

Figure 3.5 Example of Beam for Incompatible Element Test

Table 3.4 *Results of Incompatible Element Comparisons*

El Type	Distortion?	det$J$	Displ at A (mm)	Stress at B (MPa)
HEX8I	no		.00694	383.66
HEX8I	yes	free	.00620	301.79
HEX8I	yes	constant	.00603	300.70
HEX8	no		.00488	297.24
HEX8	yes		.00433	241.79
HEX20	no		.00718	501.13
HEX27	no		.00744	556.06
HEX32	no		.00750	511.01
Simple Beam Theory			.00765	480.00

The results show a wide variation in the key displacement, as would be expected from the relative coarseness of the mesh. The quadratic and cubic elements show reasonable convergence, quite close to the simple beam results, although reduced integration could not be used because of mechanisms. The HEX27 displacement is better than that from HEX20, although the stress appears to be quite high. As expected, the undistorted incompatible HEX8I element gives a good improvement over the basic HEX8 element, approaching the HEX20 result. The distortions produce greater errors in both cases. The version with constant det$J$ offers no improvement in this case. The results show that the incompatible elements are very effective provided that they can be used without undue distortions.

If incompatible elements are used, careful reference to the relevant software manuals should be made to check on what restrictions are enforced. Also, special care is required to check the availability and accuracy of other load types, such as body forces and thermal loading.

3.6 Performance Comparisons of Element Types

3.6.1 The Use of Benchmark Tests

Benchmark tests are generally used to verify finite element software performance over many situations. A large number of benchmarks have been designed and reported, many from NAFEMS (q.v. the NAFEMS publication list for details), but also for example in [15]. The reports define the actual problem concisely so that finite element input can be generated and run by the reader using any suitable software. Actual meshes are often prescribed, and specific target result presented, for comparison with alternative solutions or other finite element outputs. Although benchmark tests do occasionally compare different element types, they are less commonly involved in mesh convergence studies.

Convergence tests compare the performance of different numbers and types of element for simple geometries loaded by particular applied stress gradients. The results should ideally demonstrate the effects of uniform convergence, as discussed in section 4.4. The convergence rates are shown to differ over the various element types. Distortions effects, described in chapter 5, are not usually considered. Convergence tests have frequently been used to show the convergence rate of displacements only. More meaningful tests should also present the convergence rate of the stresses. Here, the way the stresses are calculated could well be an important issue in their accuracy.

3.6.2 Examples of Element Convergence

Many tests have been devised to show the convergence rates of common element types. Some have been reported in publications such as Benchmark and other NAFEMS products. Both simple and complex problems are devised. The former would typically be restricted to a fairly low-order stress behaviour, which would therefore only tax low order element types, higher element types giving exact response. Complex problems would tax all element types and have the possible disadvantage of an unknown solution. Ideally, a set of both these problem types would be required for a really comprehensive element type comparison.

As an example of a simple 2D test, *displacement convergence* curves have been produced for the plane square under parabolic shear loading, as described in section 2.6. Both quadratic and cubic displacement quadrilaterals and triangles are compared, with subdivisions of 1, 4, 9 and 16 squares to give uniform convergence. The results are shown in figure 3.6 for the percentage error in the maximum displacement (point C in figure 2.9).

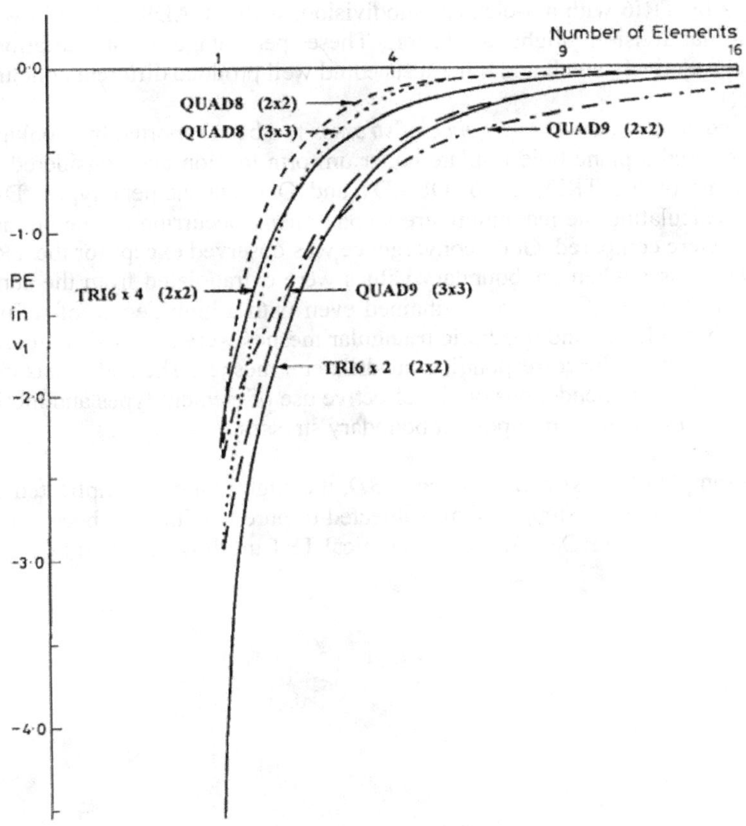

Figure 3.6 Convergence Curves for Square Shape Under Parabolic Shear

The triangles were formed by using either 2 or 4 subdivisions within each square, as indicated. Various integrating rules were used, with rules equivalent to 2x2 and 3x3 for the triangles. The loading produces quadratic stress variations, so that exact response was achieved for the cubic elements QUAD12 and TRI10 with just one subdivision, although both required at least 3x3 rules. For the other element types, all results are seen to be accurate to within 5%. The quadratic elements all behave reasonably consistently, reduced integration being beneficial for QUAD8 but not so much for QUAD9. The linear element types TRI3 and QUAD4 would have been poor performers in this test, with results for TRI3 not even appearing on the figure. Table 2.3 shows the single QUAD4 result, a percentage error of 8.2%, from a 4 element mesh, the only mesh considered for this element type.

This test shows that the cubic triangle and quadrilateral give perfect displacement results. Of the quadratic elements, QUAD8 gives the most accurate results, followed by TRI6 with a 4-element subdivision, with QUAD9 and TRI6 with a 2-element subdivision slightly inferior. These percentage error variations are, however, relatively small and other tests could well produce different conclusions.

A detailed study of the *convergence of stresses* has been reported by Honkala [16]. He considered a plane hole-in-plate under uniform tension and considered several refinements of the TRI3, TRI6, QUAD4 and QUAD8 element types. Different ways of calculating the maximum stress component, occurring on the boundary in the hole, were compared. Good convergence was observed except for the TRI3 and QUAD4 stresses when the boundary values were extrapolated from the centroidal values. Here, a significant error remained even with a high degree of refinement. Otherwise, the linear and quadratic triangular meshes were reported to give almost identical results to the corresponding quadrilateral meshes. The author has included some useful recommendations on the effective use of element types and meshes for the accurate evaluation of important boundary stresses.

As an example of *stress convergence in 3D*, the slightly more complicated case of a solid, axisymmetric, stepped shaft subjected to pure bending has been studied by R Johnson using the ROSHAZ code. A typical TET mesh is shown in figure 3.7.

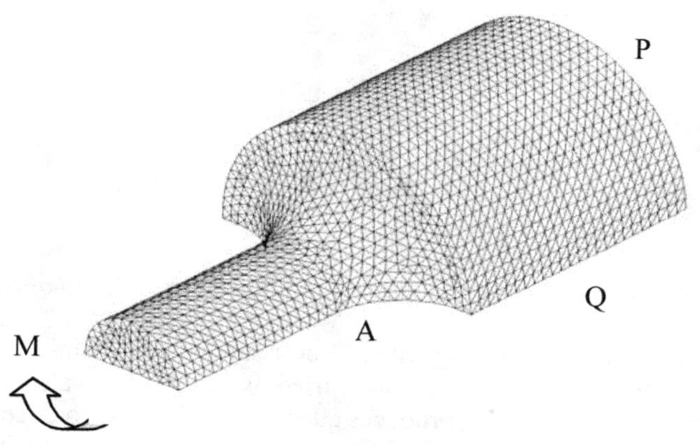

Figure 3.7 A Typical Mesh for the Stepped Shaft under Pure Bending

The element types compared for this shaft are TET4, TET10, HEX8 and HEX20, each having several mesh refinements. Two planes of symmetry exist, shown by the points P and Q, the bending being such as to produce a maximum stress (the first principal stress, or axial stress) at point A in figure 3.7. Convergence is shown in figure 3.8 for the four element types, all showing a convergence tendency to a result that is close to an approximate solution from Roark [14]. Because of the solid nature of the geometry, it was not practical to use uniform mesh sub-divisions, so selective local refinements were adopted instead. The relatively simple bending response is seen to be well captured by the quadratic elements. The HEX8 results are also quite good, the TET4 results being less so, with some oscillations. The HEX20 elements were run using 3x3x3 integration, and show good accuracy at all levels of refinement, TET10 being about the same except in the coarser meshes. Three TET10 points, shown in circles, were also obtained using the ABAQUS code, producing virtually identical stresses.

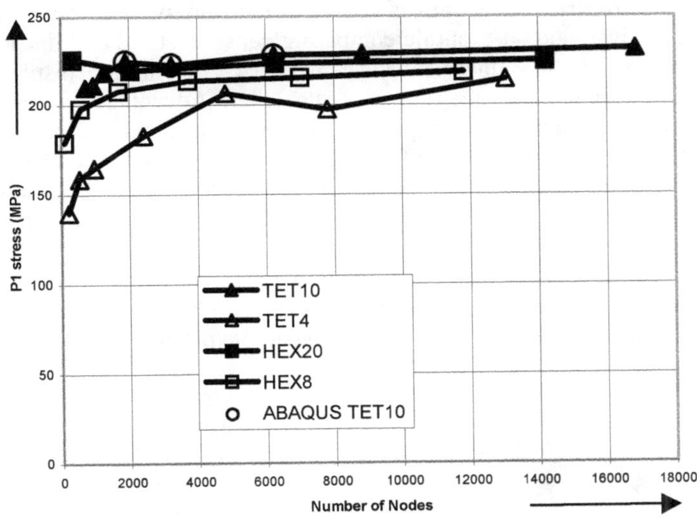

Figure 3.8 Convergence Curves for Stepped Shaft Meshes

3.6.3 The Benefits of Benchmark Tests

The above tests are used to study the performance of different element types. For expedience, the 2D tests have been chosen to be geometrically simple, but follow particular applied stress gradients. Experiences from such tests give good guidance to the mesh design of more realistic geometries, such as the described 3D test.

Such cases could be partitioned into a sequence of parts, each comprising simpler geometry and load, and being identifiable with some similar test for which convergence characteristics have been or could be determined. This would indicate a suitable mesh for each part, at least in the principal direction as discussed in the map analogy section of chapter 4. A user who has to perform analyses of new structures with no experience of adequate mesh designs would be well advised to adopt this approach.

3.7 Summary Comments

This chapter has concentrated on comparing different types of the isoparametric element families. Although the primary variables are displacements, a lot of the discussion has centred around the derived variable, the stresses, since these give more searching, and therefore more useful, information than the displacements in assessing the element performance. They are also more important in practice since most failure laws are stress based. Stress variations in a variety of different element types have been given and an example comparing constant stress and linear stress elements (equivalent to linear displacement to quadratic displacement formulations) has shown the superiority of the latter. Advantages of particular special formulations such as incompatible elements have been described and illustrated. The relative performance of several types of isoparametric element has been compared using convergence tests. The curves that have been obtained illustrate that the higher the element order, the more accurate the results for given numbers of degrees of freedom, and that reduced integration is beneficial for quadratic and hexahedral elements for many practical situations. Also, in 2D, quadratic elements perform better than triangles, although in 3D, hexahedral and tetrahedral elements frequently show a more comparable performance.

4. Mesh Design Considerations

4.1 Introduction

In this chapter, we consider various aspects of mesh design. Much of this material is covered in other "How To..." booklets, some in considerable detail such as in [5]. Here, the emphasis is on the current theme of good element usage. Various aspects of mesh design are discussed, including how much of the structure to incorporate, boundary conditions, loads, convergence, and manual and self-adaptive mesh generation. Errors and zero energy modes are considered, illustrated with examples. Finally, the basics of good mesh design are discussed using various intuitive concepts dependent on the anticipated stress or strain fields. The mesh is designed to be aligned to these expected fields, in which case element sizes, gradings, and distortions can be readily generated to produce results of competitive accuracy.

4.2 Sufficiency of Structure to be Analysed

All finite element models use meshes comprising a finite numbers of elements. The elements have to be of a particular type, depending on the structure and the loading. In reality, all structures are 3D, so that 3D elements are the most general, incorporating a complete set of field variables at every point of reference. In practice, however, structures can often be represented as 2D when the third direction has a constant thickness or is axisymmetric. The stress and strain components in that direction are invariant. When the thickness is constant, plane stress conditions are said to occur when the out-of-plane stress component is zero. Similarly, plane strain occurs when the out-of-plane strain component is zero. Such structural models offer considerable savings in total degrees of freedom, manpower and computer time.

Structures sometimes contain symmetric features, which enable **planes of symmetry** to be defined. Here, such a plane acts like a mirror and contains boundary conditions to effect this reflection. Care is required because loading and actual boundary conditions also have to be symmetric. In some conditions, anti-symmetric conditions are useful, as, for example, when the loading is represented as the sum of its symmetric and antisymmetric components. The advantage of such techniques is that the amount of structure is halved or, in some cases, quartered and sometimes even more than that.

Symmetry conditions can be specified in any finite element software. However, a more general facility is **substructuring**, which is not always available in particular codes. Here, the meshes of parts of structures can be stored in various forms for repeated use. Transformations can be applied to individual substructures to deal with geometric transformations, such as in cyclic symmetry, or to define repetitive components such as a sequence of blades around a turbine disc. Hence, the mesh of a particular component can be repeated as many times as required. The number of elements is greatly reduced and so is the number of equations to be solved, although boundary conditions and loads may have to be carefully specified.

Beams, plates and shells can be modelled with similar efficiency gains. For beams, full 3D models can be avoided by incorporating various aspects of beam theory as semi-analytical formulations in the beam elements. These include shear coefficients and moments of inertia, and require the use of nodal rotations. Plate and shell elements can include similar semi-analytical formulations, and special treatment of through-thickness behaviour, such as in plane stress. As well as being quicker with fewer equations to solve, numerical errors that would appear in 3D models with geometric thinness are avoided. However, the imposition of these semi-analytical formulations has led to many problems of finite element implementation, so that these element families have their own restrictions that need to be appreciated by the user. An indication of this lies in the large number of beam, plate and shell element formulations available in the literature, with none clearly superior to the others. In fact, research into these elements has been on-going since the 1960's.

Two areas of caution arise when using non-3D elements for beam, plate and shell type geometries. The first is when mixing different structural types of element. Examples are when a shell structure is connected to a solid 3D component, such as the welds around two mutually perpendicular cylinders at their intersection, or beams adjoining thick walls. The common nodes should ideally have the same nodal variables and be defined in the same axis system. However, extra rotational degrees of freedom in beams, plates and shells cannot be compatible with 3D elements, so either they are left to be free at the element junctions, or restrained in some way using generalised constraints.

The second is when the plate, beam or shell geometry is too complicated to be accurately represented as such by beam, plate or shell elements. Examples include intersections involving smoothing radii, small solid attachments to predominantly thin sections, and various weld profiles. The user then has to consider whether a full 3D mesh is required, when several elements may be needed to represent a particular volume that one beam, plate or shell element would otherwise have covered. The only conclusive solution is to perform both types of analysis, and compare the results. In practice, this may be prohibitive so that some well-informed engineering judgement may be required. If the plate, shell or beam model turns out to be good enough, at least repeated or similar future analyses can be conducted with confidence.

Whatever the size and complexity of the structure to be analysed, finite element modelling requires time and user effort. Hence, it should be decided at the start how much of the structure needs to be modelled, and whether details such as holes, lugs and any other local geometric variants need specific modelling. If included, they will require considerable numbers of elements locally, particularly if the features are relatively small, and will also imply large numbers of elements for compatibility with neighbouring parts of the structure. Loads and boundary conditions should also be considered when designing the model. Meshes should be finer in areas where the displacements and stresses are varying most rapidly. This is likely to occur near any geometric variants, as above, or near to where loads and boundary conditions are applied. The mesh can be coarsened away from such areas using the concept of St Venant's Principle.

4.3 Boundary and Load Considerations

Although a lot of attention should be paid to mesh generation and mesh refinement, the application of load and boundary conditions also has to be carefully considered.

Loads can be applied in a variety of forms, but are treated in the finite element theory as equivalent nodal loads. Taking pressure loading as an example, the software redistributes pressure along sides or over faces as equivalent nodal loads at each of the nodes on that side or face, usually automatically. The distribution depends on the type of element used, and is a non-trivial exercise for quadratic and higher order types of element, particularly if varying pressures are required. It is normally calculated automatically by the finite element software, but if not available, then a careful calculation of the equivalent nodal loads by the user is required. The software manuals may give some guidance on how to calculate equivalent nodal loads. The load cases used in the examples in this booklet are sometimes quite difficult to represent as equivalent nodal loads, particularly parabolic shear stresses, so care will be required if these examples are repeated.

Boundary conditions are equally important. They are usually required to constrain certain parts of the structure, or sides or faces, which are fixed in a particular direction as a line or face of symmetry. Depending on the problem, minimum fixings should be defined, since over-constraining is the equivalent to imposing additional loads, which can sometimes be relatively very large. This then produces answers to the wrong problem. At least 3 degrees of freedom need to be fixed in 2D, and 6 in 3D, to avoid floating, or arbitrary rigid body movement.

An example illustrating loads and fixings is shown in figure 4.1. Consider two simple beams under three-point bending. Beam A has simple supports at both ends and a point load P in the middle. Beam B has encastré supports at both ends with the same load. By symmetry, only half of each beam needs modelling, as shown. The right-hand edge is a line of symmetry, which requires all the displacement components in the horizontal direction to be fixed to zero, and with the load P halved. For beam A, the left-hand side only requires one point fixed vertically at the bottom. For beam B, all the left-hand side displacement components are fixed. If local stresses were required near the point load or near the beam A fixed point, a local mesh refinement in those areas would be required since both produce local singular effects. Such refinements would not be required along the left edge of beam B. The difference in fixings between A and B means that beam A can move vertically and horizontally at its left-hand side, whereas beam B cannot. Thus, beam B is equivalent to beam A with a very large force along the left-hand side restraining these vertical and horizontal movements. In more complicated real-world problems, it is quite easy to make such over-constraining mistakes.

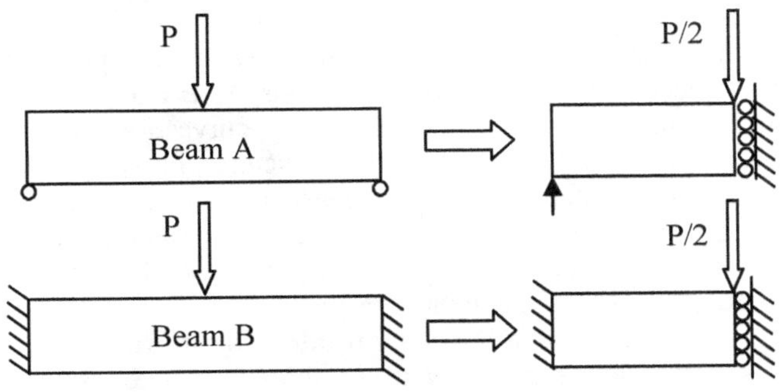

Figure 4.1 Beam with two Different Types of Support

4.4 Mesh Convergence, h and p-Type

For a given type of element, within practical limits the more elements used to model a given structure, the more accurate the results. Thus, the problem solution is said to **converge** towards the correct one as the mesh is refined, i.e. the number of nodes or degrees of freedom is increased. For simple enough structures for which a non-trivial analytical solution is available (i.e. with strain variations higher than constant), a mesh with a relatively small number of elements of a particular type may produce exact results to this solution, which we can term **perfect convergence**. Such is achieved by, for instance, a beam in tension or under constant moment, and this is the basis of the single element tests described in chapter 6. However, most structures are too complicated to have the perfect convergence property. In theory, an infinite number of elements would be required for this, which can never be achieved since computer resources limit the numbers of elements. Also, the resulting huge numbers of elements and equilibrium equations could well produce serious round-off errors in both the stiffness matrix calculations (of relatively very small elements) and in the solution of the equilibrium equations.

This progression to the correct solution with increasing numbers of elements is known as **h-convergence**. The rate of convergence for any element type is O(h^{p+1}) for primary variables (displacements) and O(h^{p}) for secondary variables (strains, stresses), although O(h^{p+1}) for optimal point stresses. Here, p is the order of the element (the shape function polynomial order) and h is a representative element size. In the example of section 3.3, $p=1$ for the linear displacement (constant stress) element, and $p=2$ for the quadratic displacement element. Table 3.3 shows that the error in stress is $1/n$ when there are n elements, equivalent to $h=1/n$ and a stress error of h, as predicted above.

For any given displacement component, figure 4.2 shows a typical convergence curve, including successive mesh refinement levels A, B and C. Most common elements will produce monotonically increasing curves, as shown, so that coarsening effects an over-stiffening of the structure. In some special cases, however, the curves are monotonically decreasing, such as when using either reduced integration or incompatible elements. The ideal refinement would produce a tolerable working error such as that produced by mesh B, depending on the accuracy required. This in turn may be influenced by how accurately the given data is known (e.g. loads) and what factors of safety exist from the engineering viewpoint. This approach ensures a sufficient although not excessive number of elements, thereby optimising user preparation time and computer resources (run-time, disc and memory space). User skill is required in producing a suitable refinement. A simple technique, known as Richardson extrapolation, can be used for an improved estimate of all the displacements from their values calculated from meshes B and C [9].

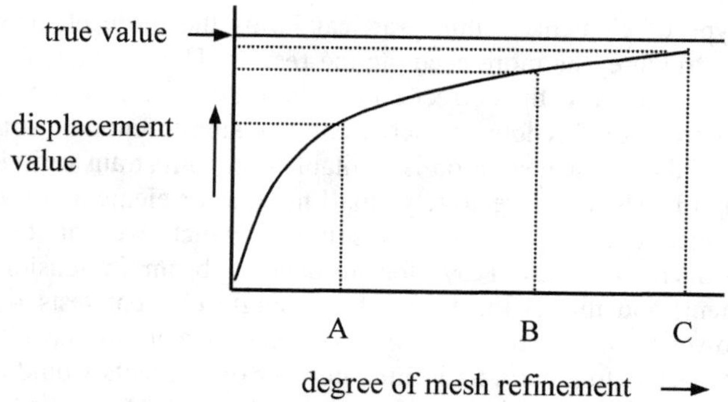

Figure 4.2 Typical Convergence Curve of a Displacement Component

The use of **mesh grading** is another effective way of reducing the total number of elements. When it is known that certain areas of the modelled structure have relatively rapid changes in displacements and stresses with respect to geometry, such as near stress concentrations, the elements should be sufficiently small. Away from such regions, larger elements will suffice. This is because, within each element, only relatively simple variations can occur, so for rapid variations, a suitably large number of elements is required. Figure 4.3 demonstrates this principle in 1D.

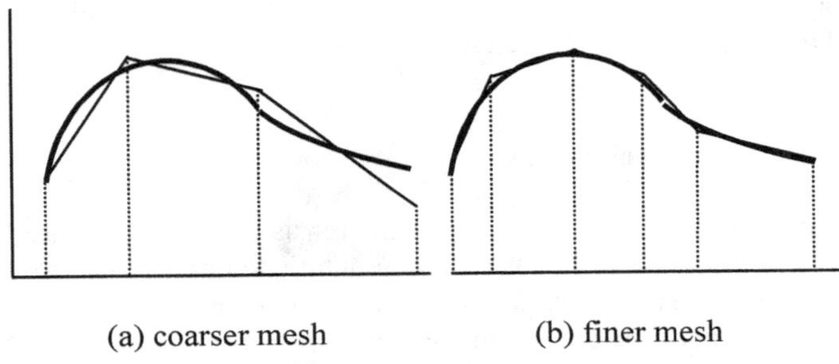

(a) coarser mesh (b) finer mesh

Figure 4.3 Element Spacing in 1D to Follow a Severe Stress Gradient

In 2D and 3D, it is inevitable that some elements will have to be somewhat distorted to fit the above gradation requirements. This can also happen when fitting elements around detailed geometric features and adjacent to curved boundaries.

The ideal mesh design, graded as above, should produce the same magnitude of results error (known as *discretisation* error) at each node of the mesh. Such ideal meshes can then be refined to reduce this error, and produce curves as in figure 4.2. Unless the perfect convergence property is available, in practice all that is required is to achieve a working error, of a few percent, at the locations of the most important results. This ideal type of mesh convergence, to the correct result for this model, is termed **uniform convergence**.

Unfortunately, the finite element method cannot furnish a single number for this discretisation error, so again some user skill in mesh design and results interpretation is important.

Another way to progress to the correct solution is not to increase the number of elements, but to increase the number of degrees of freedom, or nodes, per element. This is known as **p-convergence**. Effectively, the mesh is kept the same but the element types are changed progressively to higher shape function orders, from linear to quadratic to cubic, and so on. It is easy to demonstrate, for a given mesh, the considerable accuracy gains when going from linear to quadratic quadrilaterals.

It is convenient to slightly modify the elements to become so-called hierarchical elements [9,17]. These encompass every lower ordered element as a sub-matrix of its own stiffness matrix, which means that some of the stiffness evaluations need not be repeated during each recalculation, reducing the overall computation time. This process is easily automated (i.e. self-adaptive), and is stopped when the results improve by only sufficiently small amounts.

The advantage of this approach is that the mesh can be kept the same, and the convergence rate monitored by successive increases in the element type order. With increasing element order, distortion errors reduce. However, boundary loads and fixings require careful treatment. Also, curved boundaries would be poorly represented when starting with linear elements.

4.5 Pre-processor Mesh Techniques

Pre-processor software has been in development for as long as the finite element method itself, since it was realised early on that automated procedures to generate meshes were both feasible and desirable. User effort in generating the data for each individual element is very time consuming and error prone.

Such software is nowadays accompanied by interactive graphics, which eases the mesh generation task considerably. Standardised data interfaces such as IGES and STEP allow CAD-generated data to be transferred between different software. The difficult step then follows of assigning the topological element descriptions to fill the required volume. However, the generation of meshes, particularly for complicated 3D structures, can still be a daunting task.

Mesh generation techniques fall into two broad classes, which can be termed top-down and bottom-up.

Top-down takes for input an overall volume. This is defined to the program as a set of lines and curves in 2D, or equivalent surfaces in 3D. A mesh is automatically formed in the required volume. The problem here is that elements with badly distorted shapes can be generated to fill the space. When huge numbers of elements are being generated in complicated 3D structures, such elements can be hard to spot by eye in the graphics. Some shape checks should be available in the software, and it is prudent to use this to check such distortions.

Bottom-up is based on designing the mesh element by element, building up using sets of commands in sequence. Hence a more continuous control on each element is available. However, distortions can occur in the later stages of the process when final parts of the overall shape have to be filled. The process is also time consuming for the user.

In both cases, the meshes produced should be checked for an accurate representation of the model, and that the elements are of suitable shape and size. Some elements will have to be somewhat distorted in the fitting process, particularly around detailed geometric features and curved boundaries. The consequences of these shape distortion issues are discussed in chapter 5.

4.6 Self-Adaptive Re-Meshing Algorithms

The uncertainty of devising meshes of optimal element numbers and layout has been somewhat addressed by re-meshing using self-adaptive techniques. An initial calculation is made using an initial mesh. The results, usually stresses, are considered by software that evaluates stress "errors", representing the lack of equilibrium at each node. These errors are used to drive a re-meshing scheme, to refine in areas where these errors are large, and to coarsen where the errors are small. A second run is conducted and the stress errors are again assessed, again with a change of mesh if necessary. This process is repeated until a required accuracy is achieved. There is a tendency for good uniform convergence since the errors would tend to be roughly equal over the mesh.

Since the number of elements in the mesh is continually being altered, this is an h-convergence self-adaptive scheme. This adaptive procedure is an alternative to that described earlier for p-convergent elements. Although both are meritorious, one advantage of the h-approach is that the final mesh has well distributed elements, whereas the mesh pattern cannot deviate for the p-approach and so a bad initial mesh choice will remain.

The re-meshing algorithms work successfully for triangular and tetrahedral elements, since it is relatively easy (even for computer software) to fill 2D and 3D space with such elements. To enhance accuracy, it is also trivial to add midside nodes to create quadratic displacement elements.

In practice, a lot of elements seem to be generated, which is a particularly severe problem in 3D. If linear displacement elements are used, a large number of them will be necessary if the accurate modelling of geometric details or curved boundaries is required. The adaptive procedures are of limited use with quadrilateral and hexahedral elements, where problems of distortion would severely test any re-meshing algorithm. Thus, the advantages of these particular elements, such as reduced integration, would not be available.

4.7 Numerical Round-off and Ill-Conditioning

The finite element equations normally solve in a very accurate way. Three main kinds of error can occur:

(a) discretisation errors, due to shortfalls in the model, chiefly meshes that are too coarse and/or too distorted,

(b) round-off errors, which are produced by the solution process due to the finite size of every number stored in the computer. The round-off errors build up with continued numerical operations, and can sometimes lead to a termination in the solution,

(c) numerical integration error, if the polynomial order exceeds that permitted by the integration rule. This results in zero energy modes as discussed in the next section.

Type (a) errors should be the only ones present (obviously, it is rarely possible to have the perfect mesh!). Type (c) errors are avoided by using the appropriate numerical integration rule. Type (b) errors are normally insignificant for the types of finite element analysis that we are concerned with in this booklet. This is particularly so when the global stiffness matrix is symmetric, when, in mathematical terms, the equations are positive-definite and the Gaussian elimination process does not generate a build up of round-off errors. A good way of testing this error is to calculate the **conditioning number**, an indicator of any such build up to round-off errors. This may not be available in most commercial software. This number is calculated during the solution of the stiffness equations (2.12), and shows how many significant digits of the numbers representing the final results were lost in this calculation.

Such errors can be significant in unfavourable geometric conditions, such as when extreme aspect ratios exist, e.g. in shells, and sometimes with reduced integration.

4.8 Zero Energy Modes

In the evaluation of the stiffness matrices of those elements that require numerical integration, the rule order depends on the actual element type. Thus, for a quadratic undistorted element, it has been shown in chapter 2 that a complete rule is 3x3 (2D) and 3x3x3 (3D), or 2x2 (2D) and 2x2x2 (3D) for the reduced rule. The latter usually gives very accurate results, but occasionally zero energy modes can occur, particularly in 2D, when insufficient elements are used in a given direction.

A test for this is to multiply the total number of Gauss points in the structure by the rank of the stress-strain matrix (the number of stress components per reference point), which can be termed the **total stiffness rank**. It is over these Gauss points that the energy of the discretised system is accumulated from the stress-strain products. This rank, or number, should exceed or equal the number of **free** degrees of freedom, i.e. all the degrees of freedom less the rigid body modes or the number of constrained degrees of freedom, whichever is the greater. Failure indicates not enough contributions to the total energy have been made, and so surplus (or zero) energy modes exist. These can make the displacement results at all nodes contain arbitrary errors. If this occurs, rerun with a higher integrating rule. The possible zero energy modes, or "hourglass" modes, for QUAD4 are shown in figure 4.4, with those for QUAD8 and QUAD9 in figure 4.5. The QUAD8 and QUAD9 modes are obtained for a single element in each case, while for QUAD4, the effect is shown in 4 adjacent elements (it could also be shown in one element).

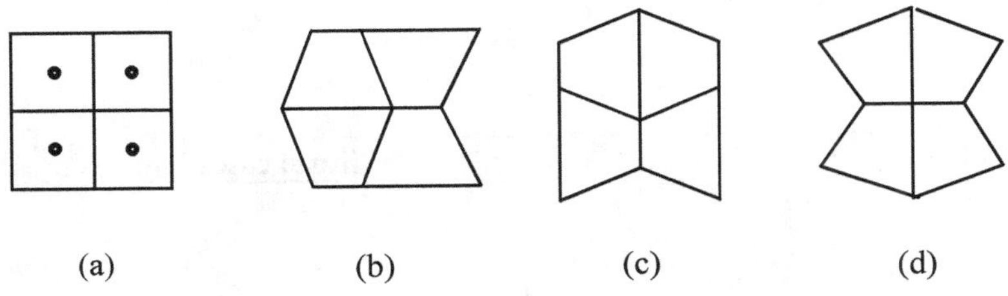

(a) (b) (c) (d)

Figure 4.4 Mesh of 4 QUAD4 Elements, Showing (a) the 1x1Gauss Points and Three Possible Hourglass Modes: (b-d)

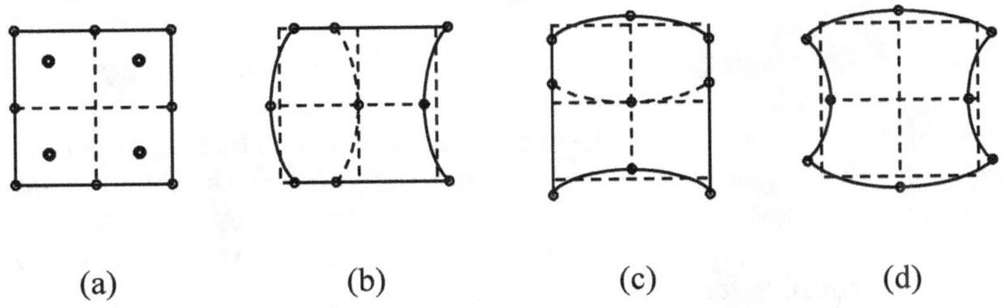

(a) (b) (c) (d)

Figure 4.5 Mesh of one QUAD8 or QUAD9 Element, Showing (a) the 2x2Gauss Points and Three Possible Hourglass Modes: (b) and (c) for QUAD9 and (d) for Both Elements

A single QUAD8 element test illustrates the above points. Using a square of unit side lengths in the cartesian directions under 2D plane stress conditions, three independent load cases are considered: (1) a constant applied x stress, (2) a constant moment, and (3) a parabolic shear, as shown in figure 4.6. Minimum fixings (three degrees of freedom) are defined at the left-hand nodes. The displacements at the right-hand nodes are shown in Table 4.1, normalised by the corresponding theoretical result or zero as relevant. They were calculated using the 3x3 integration rule The results show excellent response for all load cases, the only error of 3% appearing for the parabolic shear case, where a single element is hard pressed to model the quadratic stress variation accurately. The 2x2 rule produced very poor results for case 1, and some small but distinctive hourglass type deformations arising from the zero energy modes for cases 2 and 3, although in these last two cases v_1 was accurate to within 2%. The 2x2 displacements are not therefore included in Table 4.1. For the 3x3 rules, the nodal stresses were also inaccurate. The optimal point stresses were, however, highly accurate for cases 1 and 2, (with constant and linear stress variations, respectively) and about 10% out for the quadratic stress variations of case 3.

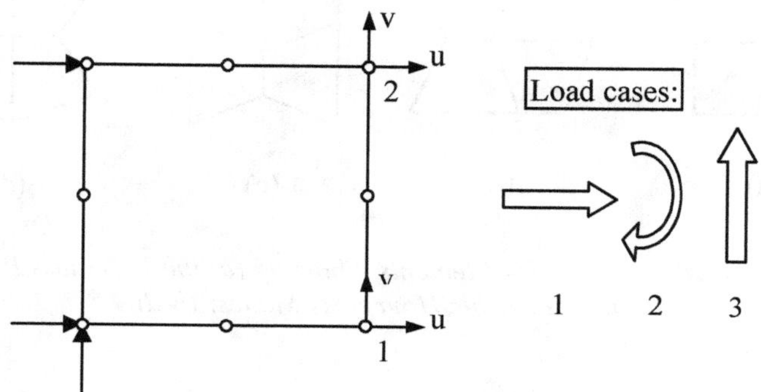

Figure 4.6 Single QUAD8 Element Test - Fixings and Loadings

For this problem, the rank of the stress-strain matrix is 3. The total stiffness rank is therefore 12 for the 2x2 and 27 for the 3x3 integration rules. The total degrees of freedom is 16, less 3 constraints, so the number of free degrees of freedom is 13. Clearly, the 3x3 results pass the test whereas the 2x2 results contain one zero energy mode, responsible for the poor performance in case 1 in particular.

Table 4.1 *Displacement Results for One-Element Test Using 3x3 Rule*

Load Case	u_1	v_1	u_2	v_2
1	1.00	1.00	1.00	1.00
2	1.00	1.00	1.00	1.00
3	1.00	0.97	1.00	0.97

Similar results can be derived for other 2D elements. Very accurate results for all three cases have been obtained when using the 3x3 rule and the element types QUAD9, QUAD12 and either 2 or 4 TRI10 triangles in the square, in each case with a repeat of the mechanisms for the 2x2 rule. High accuracy also occurs when using either 2 or 4 TRI6 elements although errors of a few percent exist for case 3 (the 4 element mesh is more accurate). For TRI6, the 2x2 rule is adequate since no zero energy modes arise. An extension of this example is given in section 6.6.

Any real structure will comprise many elements, and simple hourglass effects like those above could be difficult to spot, so the rank test is useful. However, the problem is uncommon when more than one element is used. For instance, if the above example is halved into two quadrilateral elements or quartered into four, the reduced rule now easily passes the rank test and yields excellent results. These conclusions depend on the element type. Even for the QUAD8 element, zero energy modes can appear with more than one element when, for instance, a very stiff element is pressed onto a number of softer elements as in a punch illustrated in figure 4.7. Displacements of the stiff punch can exhibit zero energy modes using the reduced rule since the softer elements offer only weak restraint to this single element, figure 4.7(a). The complete rule gives a correct response, figure 4.7(b).

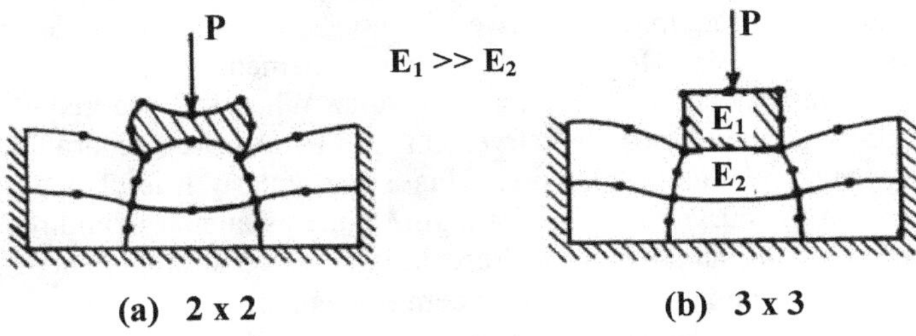

Figure 4.7 Example of Zero Energy Modes Arising from Stiff Punch on a Soft Base

4.9 Basics of Good Mesh Design

4.9.1 Elements to Fit the Geometric Shape

As discussed at the start of this chapter, the structure to be analysed should be carefully discretized and any advantages from geometric features such as planes of symmetry should be taken to reduce the extent of the model. The mesh should be such that all elements fit the space of the structure exactly, with no overlaps or voids. A good automatic mesh generator should avoid any such happenings, although it is important to check mesh plots and other diagnostics at every stage in the development. Any extra information, such as volume and weight, can also provide confirmatory checks.

The elements comprising the mesh have to fulfil two main requirements. The first is to be of suitable size and shape to fulfil the overall geometric requirements. The second is to give sufficiently accurate results. Thus, elements will need to be relatively small in areas where relatively rapid changes in displacements and stresses occur, and large where these changes are slower, to achieve the uniform convergence discussed in section 4.4. This inevitably means elements will become distorted, which can, although not always, induce errors. Hence, distortion issues are an important aspect of mesh design and are discussed in detail in chapter 5. Automatic mesh generators are hard pressed to minimise such effects.

If detailed features such as holes, attachments like lugs, and intersection radii are to be included, they will need relatively small elements to model the geometric description accurately.

4.9.2 Not Too Many Elements, Not Too Few

The ideal mesh comprises just enough elements to produce uniformly converged results of adequate-for-purpose accuracy, at all those nodes and sampling points, wherever results are required. The fewer the nodes, the less the output that has to be reviewed and stored. However, insufficient elements may cause inadequate modelling and insufficient convergence of results. When low-ordered elements are used along curved boundaries, the curves may not be followed accurately enough if too few elements exist along that side. This is particularly true of linear elements, which represent the curve as a set of straight segments and so would require many such elements to give an adequate representation. For quadratic elements, the sides are parabolic, and again give a bad fit to common curves such as circular arcs when there are insufficient element numbers. Figure 4.8 shows how parabolic sides behave over a circular arc. Cubic and higher elements match the true shape better.

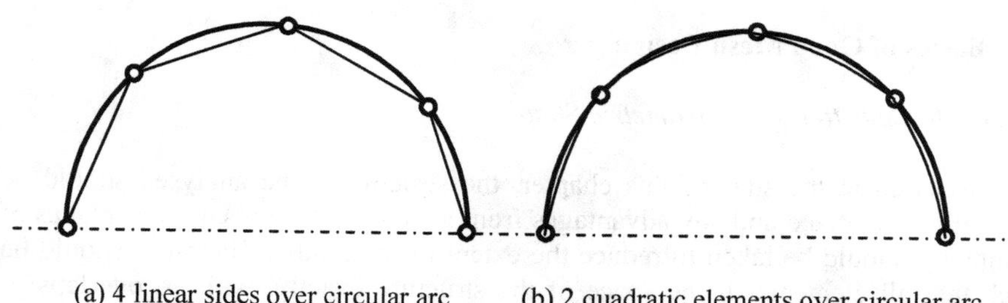

(a) 4 linear sides over circular arc (b) 2 quadratic elements over circular arc

Figure 4.8 Circular Arc Modelled by Linear and Quadratic Sides

4.9.3 Gradations to Follow Secondary Variable (Stress) Gradients

Elements need to be relatively small in areas of high stress change and large in low ones to achieve uniform convergence. This is a basic feature of mesh design that has been understood from the early days, but which has caused much anguish in practice. Many a practitioner has been told to use small elements where the stress change is high and large elements where the stress change is low. This is often inadequate advice, and a little thought will give more complete guidance.

It is easier to consider variations in the stress components rather than in those of displacement. They are usually more important, are less accurately calculated in each element, and for most elements do not exhibit equilibrium across element boundaries. A given element can only model, to the required accuracy, a given variation of stress across it. Hence, along a dominant stress gradient, the length of each element should depend on the stress gradient, as shown in figure 4.9.

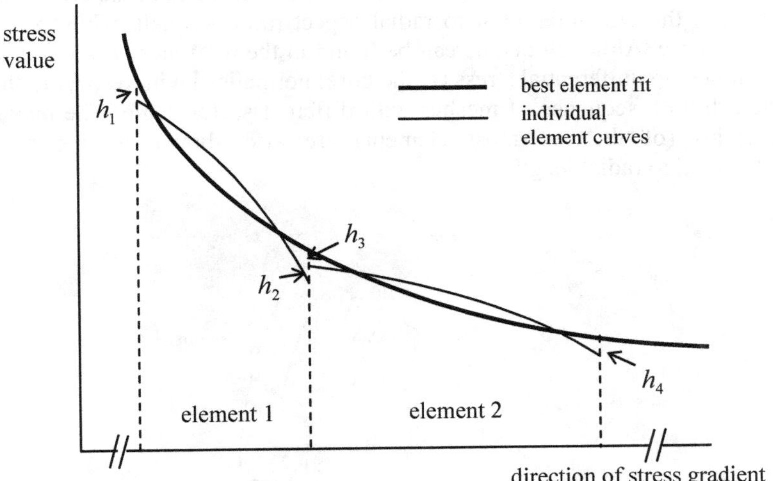

Figure 4.9 Element Modelling of Stress Gradients

The as-calculated individual stress curve (shown curved since the concept applies to any order of element) across each element is shown. Suppose the error compared to the best element fit, calculated from the optimal stress points, is h_1 and h_{2_1} at each end of the element. The neighbouring element has similar errors, and at the common nodes a good sequence of element spacings will have been achieved if these end errors h_i are all about the same in magnitude. If the mesh is further refined, the same relative sequence of element spacings shows uniform convergence that should again give similar magnitudes of these end errors, albeit of smaller amount. This value is a useful indicator of the degree of mesh refinement achieved, and is typical of the important quantities that drive self-adaptive re-meshing algorithms, as discussed earlier in this chapter.

In a 2D plane, if the stress in the direction transverse to the dominant stress gradient is constant, then element spacing in that direction can be large. Thus, it is quite permissible to employ elements with large aspect ratios. In fact, the limit on aspect ratio is only governed by the precision of the computer arithmetic. The principle applies equally in 3D. A good example is a rotating disc using 2D quadratic plane strain elements, described by Smart [18], figure 4.10. The outer radius is 4 times the inner radius, for which both the radial and circumferential stresses vary radially but are constant in the other two directions. Hence, a suitable mesh grading is only required radially as shown in figure 4.10: equal element spacing is adopted because the radial and circumferential stresses vary in different ways, with no obvious dominating gradients where the element spacing should be compacted. One element suffices axially and circumferentially, but the sector angle of the latter is varied from 0.1° to 120° to vary the aspect ratios of the elements. For small angles, the circumferential to radial aspect ratio is small, whilst for large angles it is large. Although details can be found in the reference, Table 4.2 shows the maximum circumferential stress (at the bore) normalised with respect to theory, calculated from a sequence of meshes with different sector angle. The maximum aspect ratios (of the outermost element) are also shown, as the ratio of circumferential to radial length.

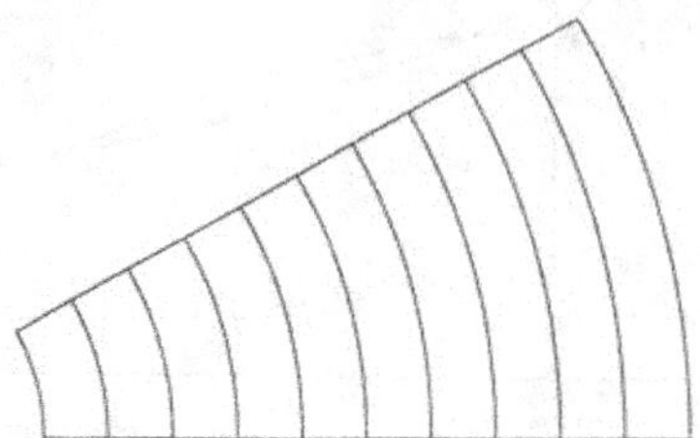

Figure 4.10 Mesh for Rotating Disc with Variable Aspect Ratios

It is seen that the stress accuracy is excellent for all angles, but as the angle increases beyond 90°, additional errors start to arise. This is because the element side is quadratic and, at such angles, is no longer a close representation of the actual circular shape of the disc, as in figure 4.8. The actual aspect ratio is not important because of the constant stress gradients.

Table 4.2 *Stresses for Increasing Sector Angles for Rotating Disk*

Sector Angle (degrees)	Normalised Stress	Aspect Ratio
0.1	1.0068	1/172
0.5	1.0068	1/34.0
1.0	1.0068	1/17.2
5.0	1.0068	1/3.4
15.0	1.0065	3.5
30.0	1.0057	7.0
60.0	1.0027	14.0
90.0	0.9968	20.9
120.0	0.9843	27.9

High aspect ratios are often useful near boundaries. Consider for instance another thick disc, this time subjected to some load system that causes the radial stress to vary significantly with radius, as shown in figure 4.11, being equal to p at the inner surface and zero at the outer surface. Because of the more rapidly varying stress values near these inner and outer surfaces, a finer mesh grading is required there compared to the more central regions, as shown. The transverse stress components are assumed to be fairly constant radially, so, whatever their relative magnitudes, do not require significant mesh refinement, as in the rotating disc. Users often place a small layer of elements near the boundary, with the next layer of elements in being much larger, presumably expecting to obtain accurate stresses on the boundary. The above arguments show that this is not a good approach since the disparity in neighbouring element sizes will give *larger* stress errors than with the element spacing shown in figure 4.11. Progressive grading is much more efficient.

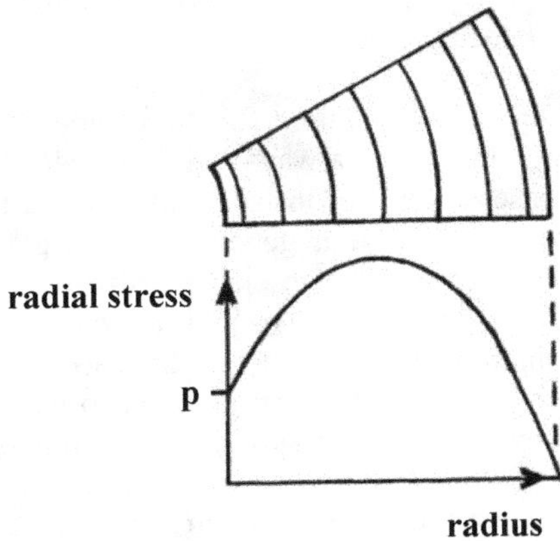

Figure 4.11 Element Layout for Pressurised Disc

4.9.4 Concept of Map Contours to Help Gradation Selection

The above developments lead to the useful concept of element design using the analogy of height contours on land maps. Thus, an isolated peak contains contours around it with spacings that depend on local slope gradients. If the peak were regarded as a dominant stress component, the contours are those of iso-stresses instead of height, and indicate how elements should be spaced radiating out from the concentration site along the lines of figure 4.9. Along the actual contours, in the transverse direction the stress component is constant, so that, in theory, no element divisions are required in that direction (although geometric considerations such as following curved boundaries may apply). Thus, a mesh can be constructed using the actual contour lines, with a required sequence of values, and, for the transverse direction, a pencil of rays with again some arbitrary angle between each. Such a mesh is developed in figure 4.12.

This simplistic approach only gives a start when a single dominant stress component exists. Unlike height, which is a scalar quantity, stress has components and the analogy implies the iso-stresses are, typically, the highest principal stress. The minimum principal stress will also need to be considered when that starts to alter due to other geometric features. In real structures, it is likely that there will be many sites where there are either stress concentrations due to geometric features such as holes and sharp radii, or a part of a boundary with prescribed loads or constraints. Hence, a more complicated pattern of iso-stress contours will arise, due to the combined effects of each of these sites, just as would happen on the land map when there are several adjacent peaks. However, the contours still give guidance to good mesh construction when using the above technique. Around the main concentration sites, the radial slopes will be uninfluenced by other sites, since the St Venant Principle applies, so the main stress gradients and resulting element spacing can be treated as before. However, transverse element spacing will have to be sized to blend in with other geometric features covered by the mesh.

This progressive build up is also required for 3D geometries, particularly for the more complicated shapes that have always been a great challenge to the finite element user. In some cases when, for instance, there are a lot of relatively small details which require modelling, mesh generators base the size of the smallest elements on these features, then keep the element size fairly constant throughout the entire mesh. Thus, if small radii, holes or lugs are to be included, the final mesh can consist of tens or hundreds of thousands of elements. This is in contrast to the computationally more efficient use of mesh grading, to use larger elements away from such details. However, this latter approach can sometimes introduce badly distorted elements and will probably be more labour-intensive. A compromise situation has to be sought, depending on the nature of the investigation.

The concept of designing meshes using the above ideas will lead to better efficiency. Elements whose shapes are formed to match the stress behaviour can be

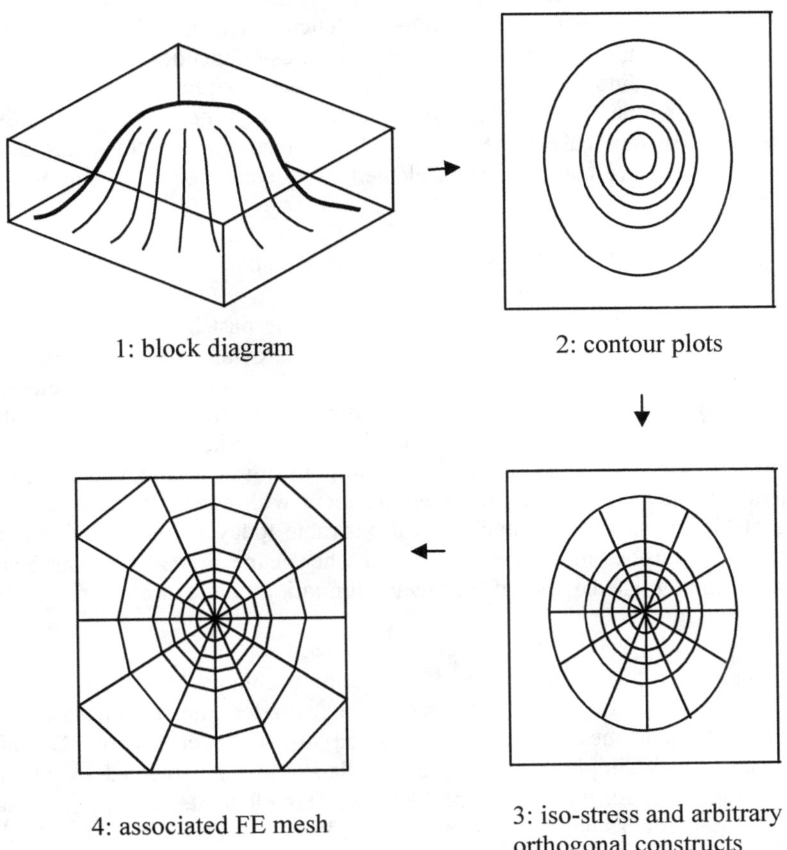

Figure 4.12 Element Meshing from Contour Analogy

termed **sympathetic** to the stress field. This field in turn depends not only on the geometry, but also on the loading and boundary conditions, so sympathetic elements are load case dependent and therefore cannot be established until some knowledge of the stress behaviour is known. Hence, a change of load case may well require another mesh design. An example of mesh dependency on loading is in pressure vessel technology, when different load cases are required. The mesh for internal pressure would require a lot less elements through the wall thickness than for a thermal transient case, when the resulting steep through-thickness stress gradients would require more layers of elements (equivalent to more contours on the land map).

4.9.5 Practical Examples of the Graded Mesh Developments

The above principles have frequently been used in the past in constructive mesh design. In the first decades of finite element use, hardware memory limitations restricted mesh sizes, so it was necessary to use carefully graded element distributions rather than simply filling the volume with an arbitrarily large number of elements. In order to establish suitable mesh refinements under those restrictions, it was necessary to consider how to mesh individual structural components efficiently, and then to assemble them with extra, carefully chosen, elements at the joins. This approach is still desirable today if the objective is an efficient, fit-for purpose mesh, particularly so when using graded meshing based on the map analogy, even though hardware limitations have been considerably eased.

As an example of this approach, a sequence of problems based on common cylindrical-type components is considered. With cylinder intersection studies in mind, basic cylinder meshes were investigated in some very early 3D finite element analyses. A simple, relatively thin-walled cylinder subjected to internal pressure is almost a constant stress problem, so few elements are required, and these are sympathetic, as defined in section 4.9.4. With quadratic elements (2D, 3D or shells), the geometric curve-fitting problem illustrated in figure 4.8 dictates the number of elements required circumferentially. One element around 180° is clearly too distorted from the true curve shape (figure 4.13), but results using two or more are good.

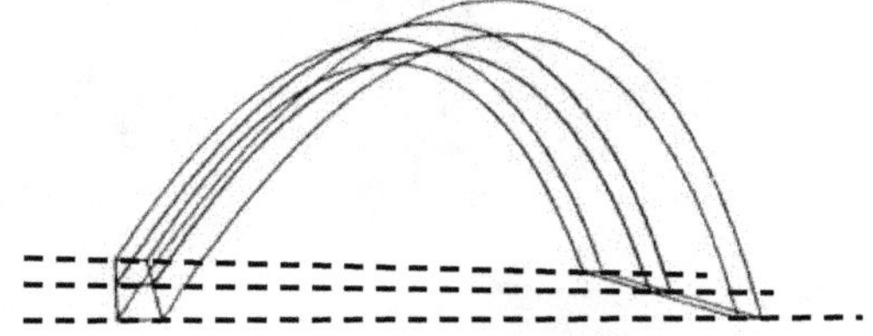

Figure 4.13 Single Circumferential Element around a Half-Cylinder

If there is no axial stress variation, only one layer of elements suffices in that direction, whatever the axial length. Otherwise, one has to establish a suitable axial element distribution. Many non-trivial axial stress states can be represented by some circumferential effect, such as a pinching collar or a fixed end. A cylinder with a fixed end and internal pressure is shown in figure 4.14. Here, 6 circumferential elements around 180° are chosen, but an axial refinement is necessary because of the rapid oscillatory change in stress approaching the fixed position. The axial element gradation is important since if it is too coarse, the resulting stresses will show considerable errors, equivalent to too few elements along steep stress gradients.

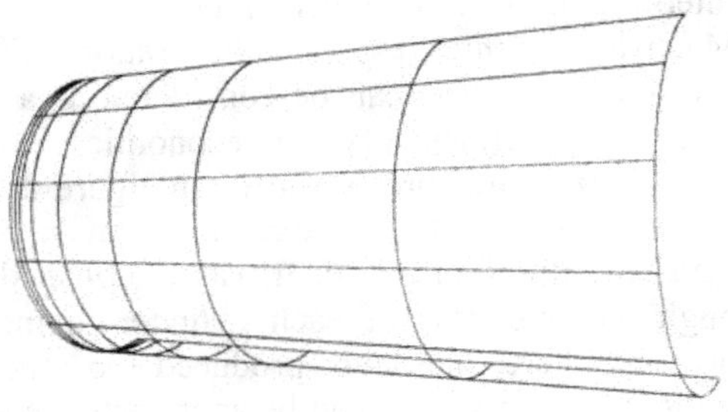

Figure 4.14 Mesh for Cylinder with Fixed End

Based on this experience, a simple extrapolation to cooling tower meshes may be made. Such a mesh is shown in figure 4.15

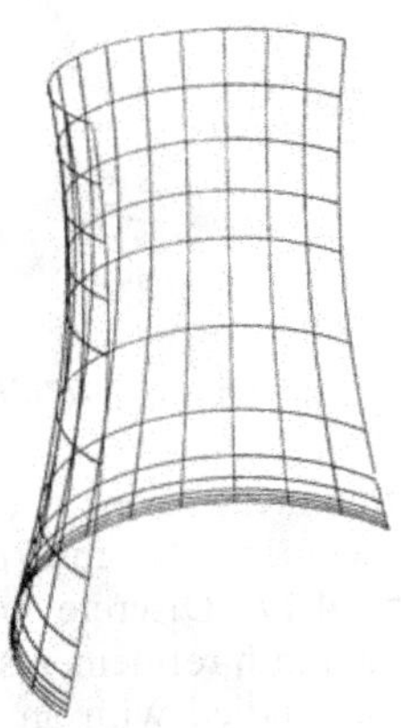

Figure 4.15 Mesh for Cooling Tower

This mesh produces accurate results when wind loading is applied, this being a very variable circumferential pressure, hence the need for 12 circumferential elements. However, the axial refinement near the bottom is far more important in order to capture the base fixing effects. This requires very short axial element lengths. Results are not impaired despite the high aspect ratios there. These meshes use either quadratic displacement shell elements (such as Ahmad or semiloof) or 3D HEX20 elements. The latter perform well even with the apparently large aspect ratios due to the thinness. For thicker shells, or when solid attachments become important, the 3D elements become necessary, hence their use in the above investigations.

Such a case is the intersection of two cylinders, representing important problems in many industries. Much higher thickness to radius ratios exist compared to the above cylinders. The junction region can be considered as a fixing, from which both cylinders require axial refinements. An economic mesh that gives good working accuracy for internal pressure is shown in figure 4.16. Elements away from the junction region are seen to be larger and can be somewhat distorted because the stress gradients there are mild. It was established that two elements were required through the thickness of each cylinder to model accurately the bending effects that occur there: one layer produced too much error while little gain occurred using three elements. This can be appreciated since the Lamé theory for thick cylinders predicts a $1/r^2$ through-thickness stress variation (r being the radius), hence the need for two (at least) linear stress elements like HEX20.

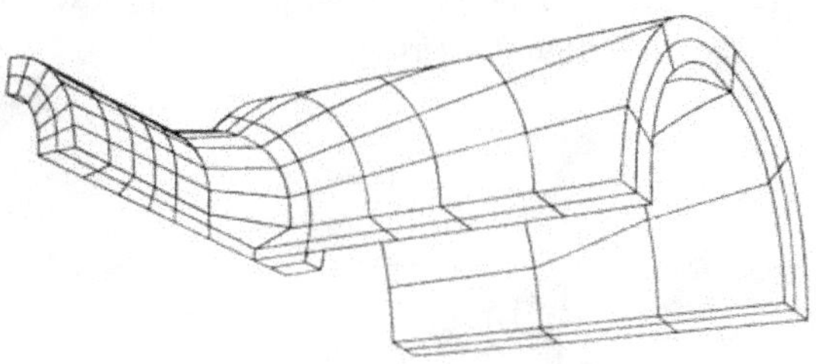

Figure 4.16 Coarser Cylinder-Cylinder Mesh

A more rigorous investigation of the intersection, when more detailed effects are to be studied, would require more elements because of the severity of the stress gradients there, e.g. as in figure 4.17. Discrete cracks, for instance, may be introduced surrounded by very local mesh refinements in that region. Alternatively, multi-intersections may need to be studied with an appropriate extension to the meshing concept, as in figure 4.18. Note that these meshes are designed for pressure loading, which gives rise to particular stress fields. The important case of thermal loading, which often arises in the investigations of these geometries, would require additional, carefully graded layers of elements through the thickness, because of the more complicated stress gradients that would then exist.

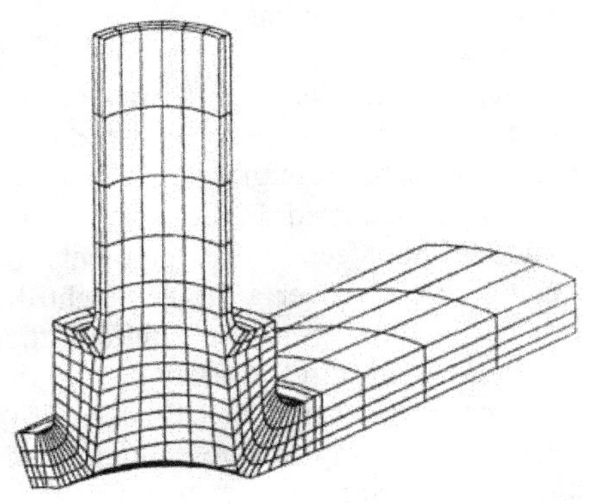

Figure 4.17 Finer Cylinder-Cylinder Mesh

Note that, in all these generic meshes, the preferred shape of each element is a piece of curved cylinder or shell, sympathetic to the geometry and the actual stress states there. It would be entirely inappropriate to model such geometries using cubic-shaped, undistorted elements.

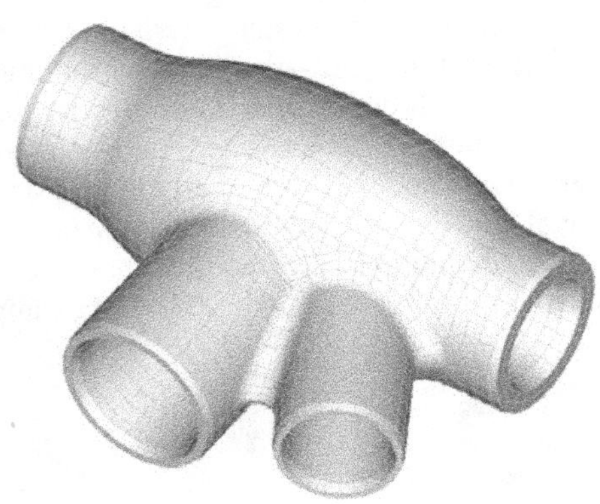

Figure 4.18 Multi-Cylinder Intersection Mesh

The above meshes have used quadratic hexahedral elements such as HEX20. They are very commonly used because of their optimum performance. Tetrahedral elements are also often used, particularly since they are easier to generate by solid modellers, although their visualisation is not so easy and they are not as accurate for corresponding degrees of mesh refinement.

A final example of the map analogy is well-known in fracture analysis. The presence of a crack tip causes singular stresses there, with local high stress gradients that indicate radially refined mesh grading. This grading does not have to be quite so fine if special elements are used. In fact it can be approximately halved: such special elements model the steep stress gradients better than ordinary elements, analogous to being able to cover a larger height difference in the map analogy. If mixed mode behaviour is to be studied, then some degree of circumferential element grading in circles around the tip is also required. Otherwise, this grading can be fairly coarse, as can areas of the mesh away from the tip region unless other stress raising features dictate otherwise. Examples of graded meshes with minimal numbers of elements for 2D crack problems are shown in a companion “How To…” booklet [19].

4.10 Summary Comments

Various aspects of mesh design have been reviewed based on the theme of good element usage. The first part of the chapter has dealt with the more traditional meshing aspects, such as the need for efficient structural selection, and suitable load and boundary conditions. Then, different aspects of mesh refinement have been discussed, including h and p-type convergence, sensible mesh grading, uniform convergence, round-off errors and zero energy modes, along with automated procedures offered by self-adaptive re-meshing algorithms. Useful concepts in achieving good mesh design have been described, based on the use of well graded elements, refined in areas where stress gradients are high and coarsened where the gradients are low. An overall impression of the required mesh has been suggested analogous to the height contours on maps, showing where element refinements and orientations are required based on the expected stress results. Some of these concepts are illustrated by several example meshes.

5. On Element Shape Sensitivity

5.1 Introduction

This chapter takes a more detailed look at how elements are used, notionally, to represent the space of the required structure. Geometrical considerations mean that individual elements take on shapes that will require distortion from the fundamental shapes considered in chapter 2. Also, element sizes will vary over the structure to achieve efficient meshing to match the anticipated stress gradients. These issues affect accuracy and, although no hard and fast theoretical rules exist, a certain amount of knowledge for good practice exists, which forms the basis for the following pages.

5.2 The Price of Meshing the Required Volume

All finite element models use meshes comprising a finite numbers of elements. Nearly all structural shapes are sufficiently complex to require that elements within them will somewhere take on some distortion from their basic shape (or *fundamental shape* of chapter 2). In many cases, curved sides or faces exist and quadratic or higher order elements are able to profitably use their curvatures. The advantages of quadrilateral (2D) and hexahedral (3D) families over triangles and tetrahedra have been described in earlier chapters, although the latter types can be made to fit more easily into given volumes, as used for instance in self-adaptive re-meshing algorithms. Whatever elements are required, a mesh generator is often used so that the user does not have full control over how the elements are formed. Indeed, because of busy work schedules and the processing of large numbers of elements, the user may not be too concerned over these issues. The software may well give diagnostic indications of elements that are excessively distorted. The danger of distortions is that those elements affected can introduce additional errors into the solution process, depending very much on the element type, the local stress field variations, and the nature and severity of the distortion. Such effects tend to be less severe the higher the order of the element.

It is prudent to check such diagnostics. In this chapter, various issues concerning distortions, or shape sensitivity, are discussed. Although some aspects of the subject still require further research, enough is known such that a given mesh can be checked out and any bad points highlighted by the informed user.

The advantages of quadrilateral (2D) and hexahedral (3D) families over triangles and tetrahedra are due to the slightly higher order of stresses that render them super convergent, and also to the beneficial effects of reduced integration. These effects are mainly obtained when using quadratic displacement, pseudo-quadratic stress elements, but are also true to a lesser extent for the linear and cubic versions and even higher orders. Where sufficient knowledge of the loaded structure's behaviour exists, the analogy of section 4.9 about using a contour map to establish the main locations of element sides or faces will go a long way to indicating good element shapes. The element distortions should then be such that the stress field variations within each element are fairly low-ordered and not error inducing.

5.3 Element Distortion Measures

Distortion measures are most important for the quadrilateral elements in 2D and their 3D equivalents. Hence we consider the basic 8-noded quadrilateral, QUAD8. The fundamental shape in which the theory is developed is shown in figure 2.6. It is a double-unit square, described in the theory space by (ξ,η) coordinates. The actual user space has cartesian (x,y) coordinates. Any change in shape in the user space from the original square is called a **distortion**. Several investigators have described the various shapes and their significance, e.g. Burrows [20], Robinson [21-23] and the references therein. A set of such distorted shapes relevant to 2D is shown in figure 5.1.

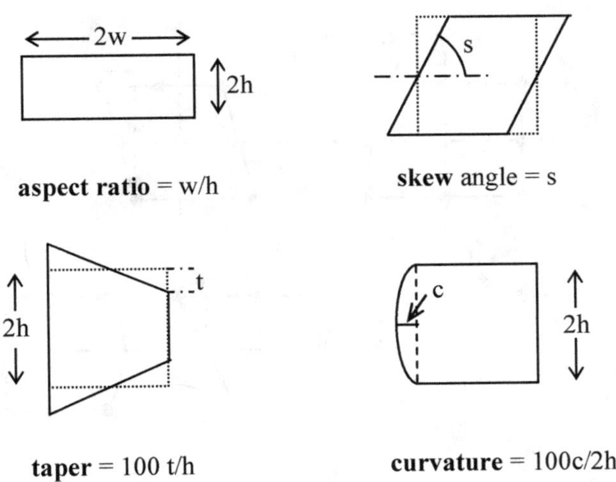

Figure 5.1 Basic Distortions for Plane Quadrilaterals

These are basic distortion shapes, which apply to other quadrilaterals as well as QUAD8, particularly the closely related QUAD9. For plate behaviour, **warping** can be added, which is an out-of-plane twist. In practice, most distortions are a combination of these. To study how they behave for QUAD8, it is convenient to divide the shapes into 5 groups. Groups 1 to 4 follow the convention of Barlow

[24] and are shown in figure 5.2, while group 5 is covered by figure 5.3. Each group has geometric distortions in the user space, which contain certain terms from the (ξ, η) polynomial in the theory space. The higher the group, the more terms in this polynomial to represent the cartesian geometry. In ascending order of distortion severity, the groups are:

- group 1: constant (ξ, η) terms, a square shape,
- group 2: linear (ξ, η) terms, a rectangle or a parallelopiped,
- group 3: quadratic (ξ, η) terms, displaced midside nodes, tapered rectangles, simple curved sides,
- group 4: quasi-cubic (ξ, η) terms, more complicated distortion shapes,
- group 5: extreme distortions, where singularities exist such that the Jacobian determinant (section 2.7), det*J,* is zero at one or more points.

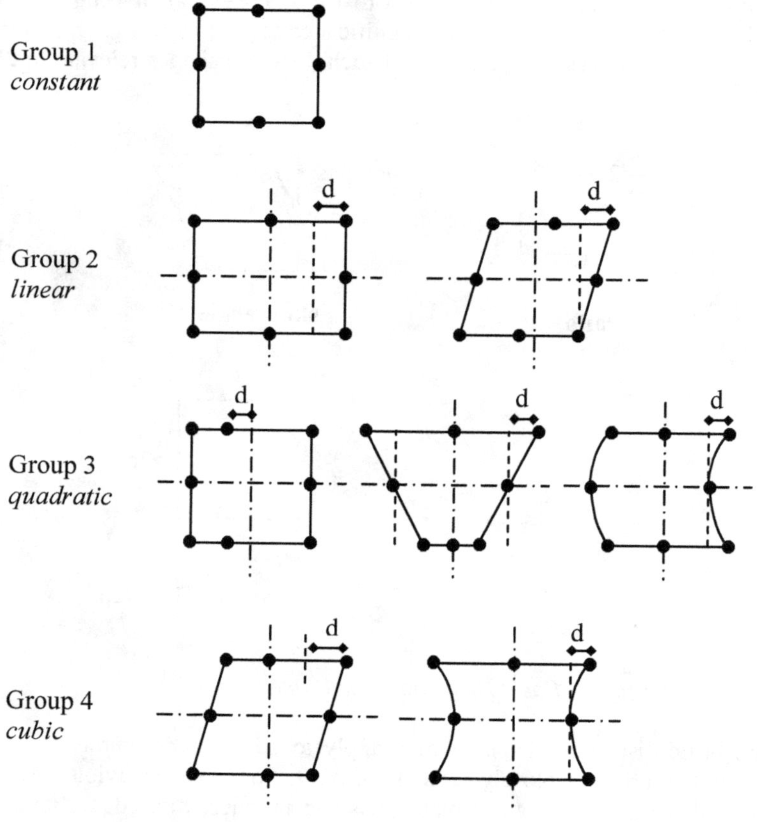

Figure 5.2 QUAD8 Distortion Groups

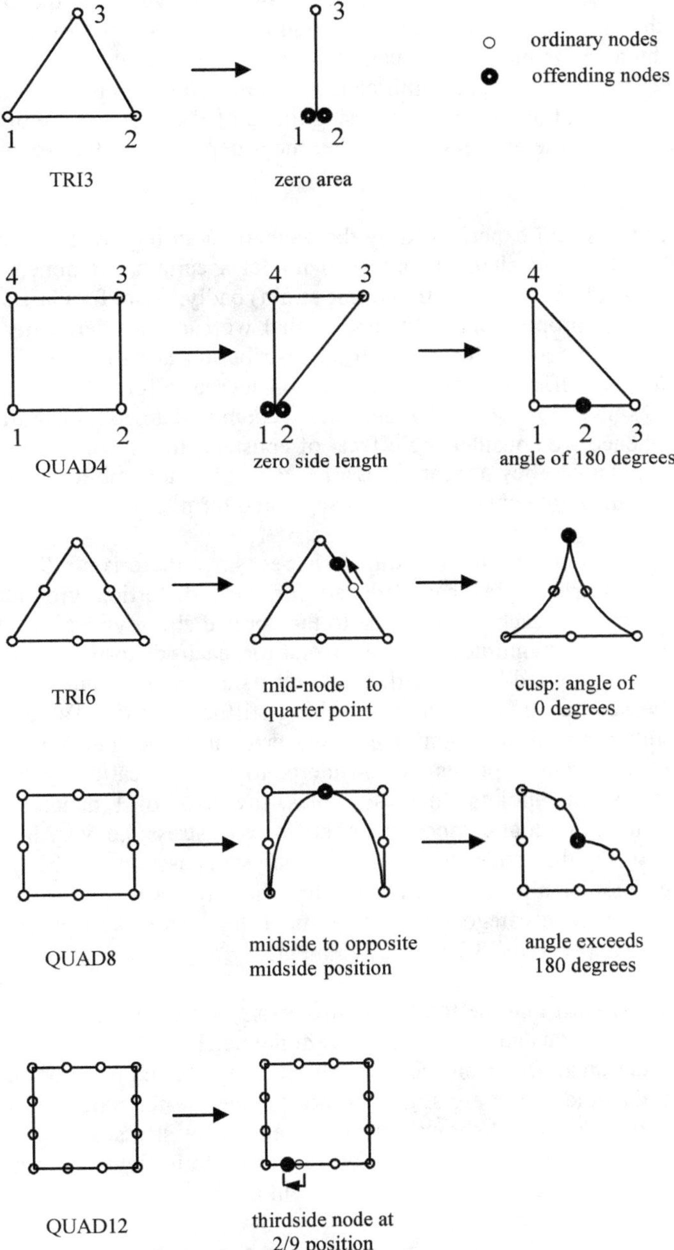

Figure 5.3 Types of Extreme Distortions in some 2D Elements

Most practical distortions will be combinations of the above, when the grouping is decided by the most severe component. In figure 5.2, the amount of the distortion is indicated by a single measure of magnitude d. Note that as the length of the side is 1, then $d<1$ and so $d^2<d$. The significance of these groupings is that errors due to the distortions depend on the group, the magnitude of d and the nature of the strain field experienced by the element. The higher the group number, the worse the error can be.

The nature of the strain experienced by the element is an important consideration, since a highly distorted element may well model a certain component of strain accurately and another (such as a transverse strain) badly. Also, for elastic isotropic elements, strain is proportional to stress, so that we can consider either strain or stress in these arguments. The exact strain distribution across an element is not known *a priori*, so different strain fields are considered in turn. The highest strain field that a single QUAD8 element can accommodate is quadratic (albeit incomplete), hence we consider the effects of constant, linear and quadratic strain variations. In practice, they appear across the element as a combined effect, but in this assessment they are considered in this separated form.

Distortion errors do not occur for group 1 shapes, since there is no distortion. For groups 2 to 4, Barlow [24] has deduced for each distortion group the error variations occurring in each element due to the applied strain variations. Table 5.1 shows the order of magnitude of the errors for each distortion group, under different orders of applied strain field. These errors are the combined errors of the calculated stresses (or strains) and the resulting stiffness matrix. Table 5.2 shows the corresponding orders of magnitude of the error in the calculated stresses (or strains) only. Here, general points are considered to be any locations in the element other than the optimal points. In these tables, the error of 1 denotes the same magnitude as the individual components of stiffness or stress, i.e. very large, whilst d is the order of the distortion, which in this analysis is assumed to be small, with $d^2<<d$ being much smaller still. In each entry, the error is for the worst case, so that some cases in each category will show much less error. As well as QUAD8, these errors also apply to the 3D HEX20 element.

In Table 5.1, it is seen that the total response errors due to any order of distortion are negligible under constant strain fields (as in the patch test, chapter 6), but are of order unity under quadratic strain fields, for both complete and reduced integration. The linear strain field errors are slightly more favourable for reduced integration, mainly of order d. Table 5.2 shows very small errors for stresses evaluated at the optimal points, never more than d even for quadratic strain fields. At other points, the errors are of order unity under quadratic strain fields.

The above groups have distortions measured by the geometric quantity d for each particular shape. An alternative way to view element distortion is the Jacobian determinant det$J$, as discussed in section 2.7. For groups 1 and 2, det$J$ is constant over the element. For the other groups, it is not constant and has been observed to vary, the more so for bigger distortions (*although not always so*). Hence, the ratio of the maximum to minimum values of det$J$ over an element gives an *a priori* guide to distortion. Many points over the element need to be sampled, particularly on sides and faces; just sampling at the Gauss points is not sufficient. The ratio is thus easily calculated for the mesh at generation time and is available for diagnostic warnings in some commercial software. A value of typically 2 is often taken as a guide for the upper limit of this ratio, such that when exceeded the distortion may be too severe, and the user should check this part of the mesh.

It should be emphasised that, in real life, some applied load cases may only give rise to insignificant distortion errors in elements where the det$J$ ratio is high, and, conversely, significant errors where the ratio is low. Evaluating the ratio therefore has to be considered only as an indicative *a priori* guide to distortion

Table 5.1 *Total Response Errors from Distortions for QUAD8 Elements*

	2 x 2 Rule			3 x 3 Rule		
	Distortion Group:					
Order of Strain:	*2*	*3*	*4*	*2*	*3*	*4*
constant	0	0	0	0	0	0
linear	0	d^2	d	0	d	d
quadratic	1	1	1	1	1	1

Table 5.2 *Stress Calculation Errors from Distortions for QUAD8 Elements*

	General Points			Optimal 2 x 2 Points		
	Distortion Group:					
Order of Strain:	*2*	*3*	*4*	*2*	*3*	*4*
constant	0	0	0	0	*0*	*0*
linear	0	d	d	0	d^2	d
quadratic	1	1	1	0	d	d

Distortion effects can be assessed by calculating the individual distortion shapes of figure 5.1 for each element. These can be checked, along with det$J$, by automatic software procedures, establishing which group is relevant to each element. Then, knowledge of the amount of distortion in each element, by diagnostic warnings of elements in certain of the above groups, or by the above det$J$ ratios, enables suitable warnings to be highlighted. The effects are, however, very dependent on the applied loading, so that particular caution would be required if multiple load cases were being run concurrently. Generally, if the contour map concept is followed, distortions following the stress gradients as described in chapter 4 (section 4.9) should be beneficial.

Group 5 encompasses **extreme distortions** involving zero values of det$J$ at one or more points in the element. Hence, the concept of using a simple measure like d is irrelevant. These have to be understood by the finite element user, and so are described in a separate section, 5.9.

5.4 Summary of Behaviour in 8-noded Quadrilaterals (QUAD8)

In real applications, the strain variations over an element are not known, but would consist of a combination of the constant, linear and quadratic strain distributions considered above. Hence in the error estimations, it is important to realise that the quadratic part has any constant and linear part removed, and the linear part has any constant part removed. Thus, the quadratic part of strain could be quite small compared to the total strain at each point, and so that error contribution, as indicated in Tables 5.1 and 5.2, may not be very severe.

Tables 5.1 and 5.2 show that no distortion errors arise when the components of strain are constant, whatever the amount of distortion. This is in fact a condition of convergence, or of passing the patch test as described in chapter 6. Hence, distortion errors only arise when linear or quadratic strain components exist over the element.

When the strain components are linear, there are no distortion errors for group 2 distortions, i.e. rectangles or parallelopipeds. This includes elements which are rectangles with high aspect ratios. The only consideration with aspect ratio is that the strain variation in each direction should be restricted in magnitude, so that short element side lengths would be required where the changes in strain are relatively large, but longer lengths would suffice transversely if changes in strain are relatively small in that direction.

When the strain components are quadratic, the distortion errors can become much larger, O(1), than when they are linear. This may not be significant if the strain magnitude is smaller, as usually occurs away from stress concentrations, but near those sites it is prudent to limit the strain range over each element. This is equivalent to having contours close together in the map analogy. Such a large quadratic strain variation would have to be reduced by subdividing into two or more elements in that direction. The behaviour would then be more like a sequence of linear segments, and the quadratic error contribution would be reduced.

As well as reducing the magnitude of the strain range over an element, successive mesh refinement also has the effect of reducing the severity of the geometric distortion in an element. An example of this is illustrated by taking an arbitrary straight-sided quadrilateral and dividing it up into four elements by joining up the opposite midside nodes, figure 5.4(a). The shape, which is in group 3 above, tends to that of a parallelopiped of group 2 as the elements get smaller. The same applies when midside nodes are not in the proper position, as shown in figure 5.4(b). An arbitrary curved-sided quadrilateral would also subdivide up eventually into a lot of parallelopipeds and so lose distortion severity. These conclusions hold for all element types in 2D and 3D, including triangles and tetrahedra, as well as QUAD8.

All strain and stress components are at their most accurate at the 2x2 optimal points in every 2D element. Their values at the nodes, if calculated directly from the nodal shape functions, will be erroneous, particularly in regions of high stress gradients where the results are probably required. Although nodal stress averaging helps to reduce such errors, a better scheme, if available, is to extrapolate the optimal point stresses out to the nodes (as indicated in figure 7.1) and use them after averaging over the adjacent elements. A useful guide to the stress errors is to compare the components calculated using the nodal shape function with the extrapolated optimal point components. The difference could be large in areas of high stress gradients, and is due to the quadratic part of the element stress field. It could be used as an indicator to add in more elements locally, as used in self-adaptive re-meshing algorithms.

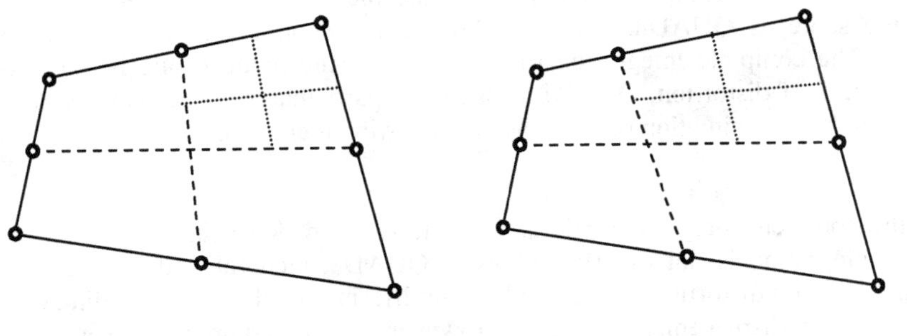

(a) side nodes in proper position (b) one side node not in proper position

Figure 5.4 Subdivision of an Arbitrary Quadrilateral

The numerical evaluation of the element stiffness matrix is seen in Table 5.1 to have similar distortion errors for both the complete integration rule (3x3) and the reduced rule (2x2). The errors can be large for all shapes under quadratic strain, which again implies sensible mesh refinements in regions of high stress gradients. For linear strains and quadratic distortions, the reduced rule is actually better than the complete rule for the overall solution (stiffness) and the optimal point stresses. This is observed frequently when using such elements to model curved shells and beams under constant pressures or moments, and as discussed in chapter 2.

5.5 Distortion in Other Quadrilaterals

Much of the advice given for the 8-noded quadrilateral applies to other quadrilaterals, from the linear displacement element with 4 nodes to the Lagrangian 9-noded and the cubic 12-noded elements. These are considered in turn. Individual element details are given in Table 2.2 and shown in figures 2.2 and 2.3.

QUAD4
The complete integration rule is 2x2, and the reduced one is 1x1, although this produces mechanisms. Distortions can be made as in figures 5.1 and 5.3, except that midside node effects are not relevant. Generally, there will be greater errors for a given distortion than with the corresponding distortion in a QUAD8 element, so the extent of any distortions should be kept small, by increasing the mesh sub-divisions if necessary. This element has an incompatible version, described in section 3.5. Bubble functions are used to enhance the linear displacement accuracy to almost that of the quadratic elements. However, if det$J$ is not constant (i.e. like group 3 above), accuracy may be lost, requiring certain formulation changes.

QUAD9
This element has a complete quadratic displacement polynomial, and performs about the same as QUAD8, but is slightly less economic on total degrees of freedom. The complete integration rule is 3x3, and the reduced one is 2x2. For small degrees of distortion, QUAD8 is superior, particularly when using reduced integration, but this situation reverses as the distortion increases.

QUAD12
Here, the complete integration rule is 4x4, the reduced one being 3x3, although reduced integration is not as effective as in QUAD8. Generally, these elements perform well, and distortion effects are less significant than the corresponding ones in the quadratic displacement elements. However, care is required in ensuring the third-side nodes do not deviate sideways from these positions, otherwise large distortion errors can occur.

Higher Quadrilaterals
Such elements use quartic and higher displacement functions. These are commonly used in p-convergence software, and can cover larger areas of the structure since they have higher displacement and strain variations. Higher numerical integration rules are required. The elements are less susceptible to distortion errors but, because of several nodes along each side, do not give an even distribution of nodes

over the space. The side nodes have to be accurately positioned to prevent distortion errors.

5.6 Distortion in Triangles

The family of 2D triangular elements do not have the advantage of reduced integration or optimal stress locations, but compared to quadrilaterals have the advantage of filling a given area more easily, particularly in self-adaptive re-meshing algorithms. Individual element details are given in Table 2.2, and some are shown in figure 2.4. The constant stress element TRI3 is the original finite element, but unfortunately is a very poor performer, an order of magnitude more elements being required than for quadratic elements to give the same accuracy. The quadratic triangular element TRI6 has 6 nodes with a complete quadratic polynomial for the displacements. However, there are no optimal points, and so there are no obvious locations where stresses are at their most accurate. As long as the sides remain straight in any of the triangles, the scalar det$J$ is constant over the element, implying that distortion may not be a severe problem. If such a triangle is much longer in one direction than another, then a high aspect ratio will exist but its effects will be as stated above for the 8-noded quadrilateral.

5.7 Distortion in Other Situations

Any element based on isoparametric principles will exhibit distortion effects as above. This includes certain beam, plate and shell elements. Example of this are the semi-loof thin shell and Ahmad thick shell elements, [25], whose behaviour in a number of situations has been observed to be as in the related elements QUAD8 in 2D and HEX20 in 3D.

5.8 Distortion in Three Dimensions

The 3D element families again behave under distortion as their 2D counterparts. Individual element details are given in Table 2.2, and some are shown in figure 2.5. Summary information of the more important elements is:

4-noded tetrahedron TET4
This is the simplest 3D element, permitting constant stress. As in the 2D equivalent, TRI3, this element is perhaps the worst performer of all element types, requiring a very large number of elements to solve even modest structures to reasonable accuracy.

10-noded tetrahedron TET10
These elements perform in an analogous manner to the 2D TRI6 elements, although are more susceptible to distortion effects than in the 2D case. Since they permit quadratic displacement variations, they are a great improvement on the TET4 tetrahedron but are less efficient than the corresponding hexahedra.

8-noded hexahedron HEX8
This element type is the 3D equivalent of the 4-noded quadrilateral QUAD4. The complete integration rule is 2x2x2, and the reduced one is 1x1x1, although this produces mechanisms. Distortions can be made as the 3D equivalents of the QUAD4 element. Generally, there will be greater errors for a given distortion than with the corresponding distortion in a HEX20 element, so the extent of any distortions should be kept small, by increasing the mesh sub-divisions if necessary. Like QUAD4, this element has an incompatible version, described in section 3.5, but again care should be taken to minimise distortion effects.

20-noded hexahedron HEX20
This element type is the 3D equivalent of the 8-noded quadrilateral QUAD8, and as such is a very competitive overall 3D performer. The distortions can be categorised as a direct 3D extension of the 2D ones, such as the groups of figure 5.2. The complete integration rule is 3x3x3, and the reduced one is 2x2x2, the latter being very effective, as in its 2D counterpart, and coincides with the optimal points where stresses are at their most accurate.

27-noded hexahedron HEX27
As in the 2D case, this element behaves in a very similar manner to the HEX20 element. The complete integration rule is 3x3x3, and the reduced one is 2x2x2. For small degrees of distortion, HEX20 is superior, particularly when using reduced integration, but as the distortion increases HEX27 becomes more accurate.

15-noded wedge element WEDGE15
This element comprises two triangular end faces and three quadrilateral connecting faces, with midside nodes. The element is designed to be used with HEX20 elements where the triangular faces would be of the same type of benefit as 2D triangles are to quadrilaterals. This includes changing from areas of coarse to fine meshing, and around small geometric details such as re-entrant corners, small holes, crack tips, etc. The elements have the same complete integration rule, 3x3x3, and the reduced rule 2x2x2 can be used but is not quite so effective as HEX20 because of the triangular faces. However, they do not significantly diminish the reduced integration performance in a mixed mesh.

18-noded wedge element WEDGE18
This element is the Lagrangian form of WEDGE15, for use with HEX27. Its performance is similar to WEDGE15 and would be required in equivalent circumstances.

5.9 Extreme Distortions (det$J$=0)

The amounts of distortion considered above have been assumed to be relatively small compared to the element size, as typified by the magnitude of d in section 5.3. As distortions increase above these amounts, errors become more significant, then eventually a singularity situation is reached when det$J$ becomes zero at some point in the element. Further distortion will then render det$J$ negative and some adjacent area or volume of the element will also have zero or negative values of det$J$.

Such distortions are herein called **extreme distortions** and are referred to as group 5 of section 5.3. They arise in a variety of situations depending on the element type. The more obvious ones are described below, with some illustrated in figure 5.3:

- for all elements, extreme distortions exist in the extreme cases of zero area (2D) or volume (3D),
- for linear elements, extreme distortions exist for TRI3 and QUAD4 when either a corner angle is less than or equal to 0°, or greater than or equal to 180°, or one side has zero length. The extensions in 3D apply to the TET4 and HEX8 elements, using solid angle definitions,
- for quadratic elements, the above causes also apply, with the addition of cusps due to curved sides producing a corner angle less than or equal to 0°. The extra midside nodes give rise to zero det$J$ when they are positioned at the quarter point locations or nearer to the corner node. Zero det$J$ can also occur if the midside node stays on the perpendicular bisector but is moved through the element to a position on the opposite side. Similar effects apply to the 3D elements TET10, HEX15, HEX20, and HEX27.
- for cubic elements, again the above causes at the corner nodes apply. The extra thirdside nodes give rise to zero det$J$ when they are positioned at locations which are the equivalent of the above quarter point positions. These elements are more sensitive than the lower ordered elements in that the distortion errors grow more quickly as the side nodes are moved. Again, similar causes arise for the 3D element HEX32.
- Lagrangian elements such as QUAD9 have the extra internal nodes compared to the above, but otherwise have similar causes for extreme distortions. The internal nodes are best positioned automatically at generation time, so that they are placed in the correct position and so there is no danger of creating a zero det$J$.

These types of distortion give rise to singular strain and stress, since det$J$ =0. Hence, gross local errors in those quantities will result, although the displacements may not be affected much. By St Venant's Principle, components of stress and strain at other parts of the mesh will be less affected the further away they are. Contour plots would show the erroneous noise as a local effect. The damage should be particularly localised if the extreme distortion lies in areas well from any real stress concentrations.

In some cases, extreme distortions actually model certain singularities correctly, such as the elastic and elasto-plastic strain fields around sharp crack tips. For the elastic strain field, this occurs in any of the quadratic elements when the midside nodes which are adjacent to a corner node, which is identified as the crack tip, are moved to the quarter point positions towards that corner. There, det$J$ =0 and the true strain field, proportional to $1/\sqrt{r}$, is accurately modelled, r being the distance of any point from the tip. For the elasto-perfectly-plastic strain field, a similar advantageous strain field arises when a quadratic quadrilateral has one side shrunk to zero length (three nodes coalescing) at the tip position, to produce a $1/r$ singularity.

Automatic mesh generators would normally avoid generating such extreme distortions, although in some awkward geometrical shapes they could well arise. It is therefore sensible to utilise any pre-processor checks for zero det$J$. However, checking this quantity has to be made at many locations over each element, including on its boundaries, since det$J$ has a polynomial variation which is of higher order the more the distortion, and so which requires a lot of sampling points to detect trends towards zero. This check is usefully made with the ratio of maximum to minimum det$J$ over each element, which was stated in section 5.3 to give some guidance to distortion effects.

5.10 Examples of How Distortions both Generate and Reduce Solution Errors

Many examples can be devised to show the effects of distortion. Simple tests involving only a few elements are easily conducted to allow any desired distortion effects to be compared. Note, however, that simple tests may in fact be quite severe tests in that, with only a small number of elements, there may be quite large strain variations over each element, e.g. quadratic strains in Tables 5.1 and 5.2, producing quite extreme effects. This is not normally encountered in actual application meshes, where the distortion effects would be more subdued due to both the greater number of elements and significant amounts of constant and linear strain fields. However, such examples are of great educational value, and interested readers are encouraged to perform their own experiments. Some typical examples are included below.

5.10.1 Distortion Comparisons in 2D Quadrilateral Elements

Varying stress fields across distorted elements can be studied using 2D beams, since analytical solutions are available for various load cases. A long beam, modelled as one element high and several along, has been used to test distortion effects for different element types, e.g. [8,9]. Such tests are well worth devising in order to gain familiarity and to check out any points of interest. A representative test is described here, comparing the element types QUAD4, QUAD8 and QUAD9. The beam's length to height ratio is 5:1. Two elements only are used in each type, with both distorted and undistorted element shapes as shown in figure 5.5.

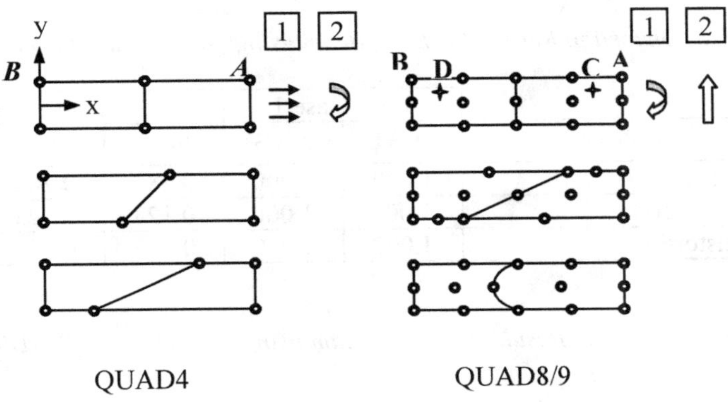

Figure 5.5 Element Shapes for 2D Quadrilateral Distortion Tests

For the QUAD4 test, 2 load cases are used. The first is a uniform stress and the second is a bending moment, both applied at the right hand end of the beam, giving responses of constant and linear stress, respectively. Minimum fixings are defined at the left hand end, 2 in the x direction and 1 in the y direction. Solutions are given in [21]. The normalised results are shown in Table 5.3, as the horizontal and vertical displacements at node A and the *x* stress component at node B (these are largest respective values). The first load case is a patch test, which is easily passed even with the distortions. The second case produces linear stress fields that cannot be accommodated by single QUAD4 elements, so meshes of many such elements are required to achieve converged results of adequate accuracy. The present results are seen to be very inaccurate. The two distorted meshes are progressively more taxing on the accuracy. For QUAD4, the one optimal point lies at the centroid, at which for load case 2 the stresses are zero and so are not presented.

The tests for QUAD8 and QUAD9 are shown in Table 5.4. Since the elements are one degree higher than QUAD4, the two load cases are chosen to also be one degree higher. Case 1 is therefore a constant moment and case 2 is a parabolic shear (linear moment), both applied at the right-hand end of the beam and giving responses of linear and quadratic strain and stress, respectively. Fixings are as the QUAD4 tests [21]. The results are shown in Table 5.4, as the vertical displacements at node A, the nodal *x* stress component SX calculated at nodes A and B, and the Gauss point stress components GX calculated at the points C and D of figure 5.5. Both complete and reduced integration was used for the stiffness calculation (3x3 and 2x2), and with Gauss points being at the 2x2 optimal locations. For the QUAD9 elements, however, the reduced rule produced noticeable hourglassing since the zero energy mode test of section 4.8 is not passed with two such elements, so the results are excluded.

Table 5.3 *Distortion Results for Long Beam using QUAD4 and Two Load Cases*

Mesh	Load Case 1		Load Case 2		
	u(A)	σ_x(B)	u(A)	v(A)	σ_x(A)
No Distortion	1.000	1.000	0.280	0.280	0.299
Smaller distortion	1.000	1.000	0.123	0.123	0.143
Larger distortion	1.000	1.000	0.089	0.089	0.210

Table 5.4 *Distortion Results for Long Beam using QUAD8 and QUAD9 and Two Load Cases*

Element	Rule	Load Case 1			Load Case 2		
		v(A)	SX(A)	GX(C)	v(A)	SX(B)	GX(D)
Undistorted Mesh							
QUAD8	2x2	1.000	1.000	1.000	0.998	1.000	1.000
QUAD8	3x3	1.000	1.000	1.000	0.977	0.915	0.945
QUAD9	3x3	1.000	1.000	1.000	0.978	0.918	0.945
Trapezoidal Distortion Mesh							
QUAD8	2x2	1.000	1.278	1.000	0.998	1.091	1.014
QUAD8	3x3	0.766	1.152	0.849	0.700	0.694	0.641
QUAD9	3x3	1.000	1.000	1.000	0.876	0.722	0.753
Curved Distortion Mesh							
QUAD8	2x2	0.827	0.800	1.082	0.805	1.271	1.053
QUAD8	3x3	0.504	0.510	0.544	0.546	0.895	0.762
QUAD9	3x3	0.751	0.977	0.805	0.750	0.844	0.922

The trapezoidal distortion belongs to group 3 of figure 5.2, whilst the curved distortion is of group 4. Table 5.2 predicts that considerable errors will occur in these cases particularly for case 2, the quadratic strain field. The results of Table 5.4 indeed show such an effect. The group 4 distortion gives the worse results, although both distortions were chosen arbitrarily and were not expected to have any comparative relationship. The det$J$ maximum/minimum ratio for the trapezoidal distortion is 3, whilst that of the curved distortion is 1.25, but the errors are greater in the latter case. Both distortions are, however, directly affecting the main active stress gradients, and so should produce significant errors, as is observed. Generally, the 2x2 results are better than the 3x3 for QUAD8, and the 3x3 QUAD9 results are better than those of QUAD8 3x3 results. The QUAD8 2x2 results are the best, followed by the QUAD9 3x3 results.

These tests show how distortions, which are not particularly excessive, can lead to large errors in results. It should be pointed out though that such distortions are deliberately chosen here to produce bad results, and are not at all sympathetic to the expected stress flows and the map analogy of section 4.9. The stress flows in these load cases are dominated by the direct x stress and the shear xy stress, so the element sides should lie in the x and y directions. The actual element spacing in those directions should correspond to the dominant stress gradients, as in the map analogy, and could be assessed by convergence tests. The present distortions were "across" these directions and therefore undesireable.

5.10.2 Distortion in a Parabolic Shear Case

Another beam, this time square, is subjected to an applied parabolic shear over one end, which produces a linear moment, linear direct stress, and parabolic shear stress fields. The effects of mesh refinement and severe element distortion for QUAD8 elements are described, in some detail to illustrate several issues.

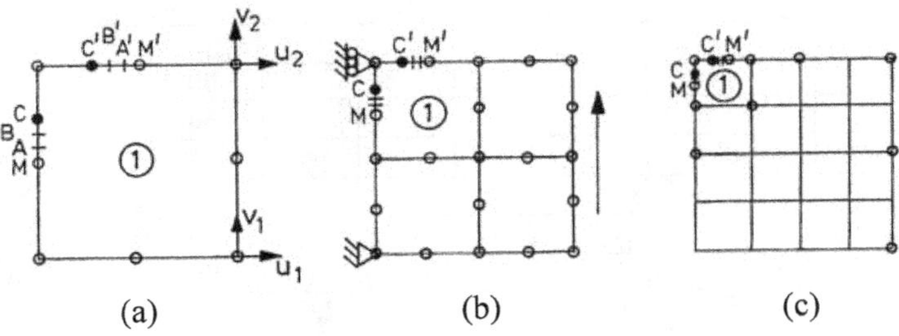

Figure 5.6 Meshes for Parabolic Shear-Loaded Square Beam

The beam is loaded on its right-hand side as shown in figure 5.6(b), with minimum boundary constraints along the left edge. An analytical solution is available [21]. Three progressively-refined meshes are considered, with 1, 4 and 16 plane stress QUAD8 elements, as in (a), (b) and (c) of figure 5.6. When all the elements are square and undistorted, excellent accuracy is obtained, so perfect convergence exists. However, to simulate increasing distortion effects, internal node movements are made so as not to disturb the external boundary shape. In particular, two nodes at the top left corner of each mesh were progressed towards the quarter point locations adjacent to the top left corner node. The positions are shown in the figures as, progressively, MM′ (undistorted), AA′, BB′, then CC′. The latter is at the quarter positions and is an extreme distortion. The AA′ and BB′ positions are severe types of distortion belonging to group 4 (cubic) of figure 5.2. Since quadratic strains exist in the elements, these meshes therefore experience the worst conditions in Tables 5.1 and 5.2.

Both complete and reduced integration rules were used, although the latter could not be applied to the single element test because of zero energy modes.

The results are shown in Table 5.5, normalised with respect to the analytical result, and include the ratio of the maximum to minimum det*J* over element 1. As this ratio increases to infinity, the results are seen to deteriorate. Even the slight distortion at AA′ gives a significant error. The results shown are the displacement components u_1, v_1, u_2 and v_2 on the right hand side of the beam, where the displacements are largest but are well away from the distorted area. This shows that the errors induced by local element distortions cannot be assumed to remain in the distorted area.

Table 5.5 *Displacement Results for Parabolic Shear Loading in a Square Beam*

Mesh	det*J* ratio max/min	2x2 rule u_1	2x2 rule v_1	2x2 rule u_2	2x2 rule v_2	3x3 rule u_1	3x3 rule v_1	3x3 rule u_2	3x3 rule v_2
(a) 1 element mesh									
Regular mesh	1					1.00	0.97	1.00	0.97
Nodes at AA′	3					0.86	1.01	1.21	1.03
Nodes at BB′	15					0.74	1.13	1.58	1.16
Nodes at CC′	∞					0.59	1.53	2.57	1.59
(b) 4 element mesh									
Regular mesh	1	1.00	1.00	1.00	1.00	1.00	1.00	1.00	1.00
Nodes at AA′	3	0.94	1.05	1.13	1.05	0.94	1.04	1.12	1.04
Nodes at BB′	15	0.87	1.15	1.37	1.14	0.88	1.12	1.31	1.12
Nodes at CC′	∞	0.81	1.42	1.97	1.41	0.82	1.33	1.77	1.32
(c) 16 element mesh									
Regular mesh	1	1.00	1.00	1.00	1.00	1.00	1.00	1.00	1.00
Nodes at AA′	3	0.97	1.05	1.12	1.05	0.97	1.04	1.11	1.04
Nodes at BB′	15	0.93	1.14	1.32	1.14	0.94	1.11	1.26	1.11
Nodes at CC′	∞	0.90	1.36	1.78	1.35	0.91	1.27	1.60	1.27

The largest error in each mesh corresponds to the CC′ case for displacement u_2. The three mesh refinements demonstrate that this largest error decreases with increasing numbers of elements, being for the meshes (a), (b) and (c): 2.57, 1.77 and 1.60, respectively, relative to theory. There are using complete integration; reduced integration is less accurate in these distortion tests. For both rules, the convergence rate is very slow and there is even a considerable error with the finest mesh use. This mesh is more than sufficient for good accuracy if these distortions were absent, and would even be suitable with higher applied stress fields. The reason for the poor results is that the chosen distortions are not sympathetic to the stress fields.

The det*J* ratio is a good *a priori* guide to distortion in this case. A value of 2 is often taken as a guide for the upper limit of this ratio, such that when exceeded the distortion may be too severe, and the user should check out this part of the mesh. In this load case, this ruling is suitable, but in different load cases, the det*J* ratio has to be considered as a guide only, since meshes and their distortions are load case dependent.

Although this test has only simple geometry, the progressive distortions cause increasingly severe effects on the results, which would be typical of those occurring in the more complicated meshes of actual applications. The beam is too compact for St Venant's Principle to apply, so all results are affected. This principle may be advantageous in more complicated geometries, such as an assemblage of beams where, if only one beam had bad distortions, the errors may well not propagate into the other beams.

5.10.3 Examples of Advantageous Distortions

Elements are distorted from their fundamental shapes usually to fit into a required space governed by the overall geometric shape and local mesh refinement, a process often automated by mesh generation software. It is very rare for distortions not to occur, but if the distortions are sympathetic to the resulting stress fields, this works to advantage. Good examples of this are in the cylindrical type problems described in section 4.9.5. There, both 2D and shell quadrilaterals and hexahedra in 3D had to take the shape of curved segments to fit the geometry, but showed that no distortional errors resulted because their shapes corresponded to the stress fields. The concept of sympathetic distortions can be applied to many generic types of structure. If new meshes are required for such cases, it is well worth investigating these effects on some simpler cases, as was done in the simple cylinder tests.

Extreme distortions are used in fracture mechanics to model the strain singularities around crack tips. The best known case is the quarter-pointing effect in quadratic elements, as described in section 5.9. The midside nodes of elements surrounding a crack tip node are all moved in to the quarter position, making det*J* zero at the tip node. A stress field occurs whereby all components are singular at the tip and exhibit the required slope across the tip elements. This considerably increases the accuracy of the calculated fracture parameters such as the stress intensity factor. Elements may be graded to become progressively larger away from the tip, [19].

5.11 Summary of Designing Meshes to Benefit from Distorted Shapes

The above investigations into the shape sensitivity and resulting distortion errors for many popular elements give guidance into good practices for practical mesh generation. Although the subject has been covered in section 4.9, some of the important issues resulting from shape sensitivity are highlighted here.

For any element, in the more highly stressed regions such as around stress concentrations, the element spacing in the direction of the steeper gradients should be small enough that the stress range over each element is limited. This is equivalent to crossing contours in the map analogy, such that an element only traverses from one contour to the next, the contour spacings having been chosen to be equivalent to the required maximum stress range per element.

Element shapes should be designed to be sympathetic to the problem in hand. Distortions should be made to fit the geometric requirements, ideally with the element sides parallel to the directions of the main strain and stress components, e.g. the principal directions, if such flows can be assessed *a priori*. For the 8-noded quadrilaterals QUAD8 and QUAD9, this should ensure that the strain variation is linear, with perhaps only a small quadratic part, so that the errors typified by Tables 5.1 and 5.2 are contained. These errors are worst cases within each category, and are probably less for such use in practice. In particular, element sides should follow accurately curved sides to ensure the correct volume of material is modelled.

Provided the mesh can be readily generated, the use of QUAD8 and QUAD9 quadrilaterals, and HEX20 and HEX27 hexahedra, are high performance elements, which offer important features such as reduced integration and optimal stress points for accuracy beyond their order. Providing stress gradients are not too high, each element can cover a reasonable amount or area, so good meshes can be devised with modest numbers of elements. The nodes are also well distribution over the space. If such meshes cannot be readily generated, due to either very complicated geometric details or the production of unavoidably large distortions, it may well be more beneficial to generate triangular or tetrahedral meshes.

Some aspects of distortion can be very beneficial to designing meshes with enhanced accuracy.

6. The Value of Single Element and Patch Tests

6.1 Introduction

The idea of using a mesh of only one element for a simple structure gives rise to the concept of the single element test. The response to a given load is used to judge how different elements respond to different orders of stress (and therefore strain) variation across the element. Only linear elastic solutions are needed to test these basic element properties. Typical structures are beams, which can be modelled in either 2D or 3D, and which are of conveniently simple geometry and can be subjected to any order of loading. This loading can be defined to give prescribed stress variations. The constant stress state is used for the **patch test**. This is a fundamental test requiring a small number of elements, and is used to evaluate the very basic performances of elements as a means to ensure that they are capable of converging to correct results with mesh refinement. Other tests with single elements are based on higher applied stress variations, such as linear or quadratic. Here, some elements can give an exact response, or **perfect convergence response** as defined in chapter 4. Thus, a quadratic displacement element gives correct results for a shear-loaded beam that contains linear stress variations. Others, such as bilinear or constant stress elements, would alone produce poor results and need mesh refinement for improvement. All these tests make comparisons of different distortion measures such as varying aspect ratios, skewness, taper and any other geometric properties that might affect element performance.

Such tests are of value because they are easy to set up and run, and yet can test out all the above features, along with any inherent instabilities that characterise the element. Indeed, under severe enough conditions, the elements will be much more severely tested than elements in large production meshes with numerous elements representing complicated field problems. Hence, the educational experience gained from simple tests can be of great value to future practical applications.

6.2 A Brief History of Benchmark Testing

Simple tests of element behaviour have been devised ever since finite element software was first produced in the 1950's. After the initial tests had shown that the solution techniques and element formulations had been implemented accurately, the engineering performance of the elements had to be assessed to convince potential users that the elements were worth using! Initial testing for this consisted of devising simple meshes for elementary problems having known analytical solutions, then comparing the results. This was followed by mesh convergence studies, whereby the number of elements was progressively increased to improve accuracy. This was particularly significant then since only linear displacement, constant stress, elements were available, with slow convergence rates. Although dedicated to elastic membranes and plates initially, the convergence test has been used ever since during the wide expansion of the finite element method.

These simple tests gave rise to the concept of **benchmarks**, which are well-defined tests involving meshes of particular element types or element families subjected to given loading, boundary conditions and materials behaviour. In order to organise the large numbers of benchmarks that were beginning to appear into well-chosen ones within each required category of finite element technology, NAFEMS, through its working groups, initiated a series of benchmark reports, each covering a related set of tests. The range now available includes elastic behaviour in membranes, solids, plates and shells, material and geometric non-linearities, dynamics, fracture, and other categories. The different aspects of analysis, including choice of element type, convergence, material properties, algorithms, etc., are specified in each set of the benchmarks as required. The relevant reports are listed in the current NAFEMS publication list [26].

Another type of test was being devised in parallel to the above, known as the concept of the **single element test**. Here, the idea was to test just one element of each type with a simple shape but with non-trivial loading. Results were compared to an analytical solution. Elements could be distorted. This contrasts the established **convergence test**, which compares different numbers of elements to test convergence rates. One benefit of the single element test is that if it behaves well as a simple structure, then so should an assemblage of these element types. The applied stress distribution across the single element could be varied so that, whatever the order of the element, a high enough variation would exist to severely test the element's response. Lower ordered variations would be easily dealt with, and should give correct results so that convergence needing many elements was not an issue. The tests are conducted over, firstly, undistorted elements, then progressive distortion can follow along with other features such as different orders of numerical integration rule.

The early tests showed surprising deficiencies in commonly used isoparametric elements, so quite a lot of activity ensued through NAFEMS. In particular, Burrows [20] conducted a large number of single element tests for 2D membranes, with three main distortion parameters, three applied stress variations and four types of element. This work is reviewed later in this chapter.

6.3 The Patch Test

This is like the single element test in that convergence is not an issue. However, a few elements are needed to test out the ability of a given element to reproduce a state of constant stress or strain [8,9].

The procedure requires a patch of elements that share at least one common interior node, as shown in figure 6.1. A state of constant stress is applied through suitable equivalent nodal loads or prescribed displacements at every boundary node, although in the former case some degrees of freedom will require minimum fixings to prevent rigid body motions. The internal nodes are neither fixed nor loaded, and it is their presence that requires more than a single element in the test. The stresses throughout the elements are computed and the patch test is passed if exact agreement with the constant state exists. Any computed differences will then only be due to round-off errors (section 4.7). It is important to define only minimum fixings, since over-constraining effectively introduces extra loading which removes the state of constant stress. Since the stresses should be constant, so also should be the strains, and this is a useful check that the latter have been implemented and computed properly.

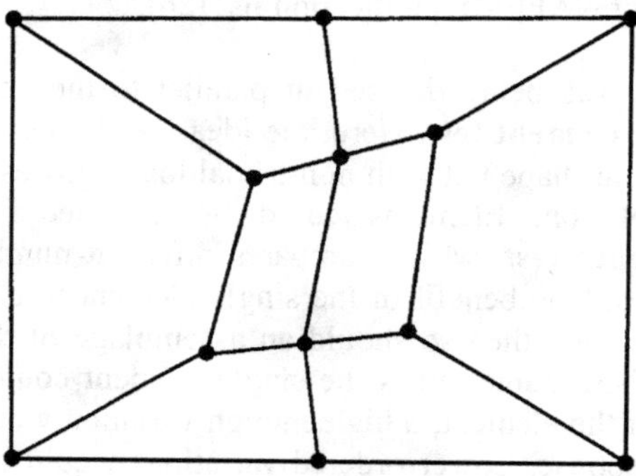

Figure 6.1 Element Layout for a Patch Test

The test should be repeated in turn for all constant stress states available to the element, such as the three components σ_x, σ_y, τ_{xy} in 2D and the six components σ_x, σ_y, σ_z, τ_{xy}, τ_{yz}, τ_{zx} in 3D. For other types of structure, the relevant constant stress states again have to be covered, such as those generated by constant bending and twisting moments for plate bending elements. The elements may have arbitrary distortions, although some extreme cases may prohibit passing the test, in which case care in using such degrees of distortion will be required.

This is a fundamental test of element behaviour. All popular elements pass this test, although reduced integration can be restrictive. Also, extreme distortions as discussed in section 5.9 may produce errors in elements which otherwise pass this test.

The ability of an element to reproduce constant stress/strain as above ensures that mesh refinement produces convergence of results to the correct solution. This is a fundamental requirement, and so the test is more of use to code developers, to ensure that element routines are implemented properly before releasing the software to the public. The user would probably only use this test for simple familiarisation with new software or perhaps in investigating special areas, such as those mentioned above when doubts might arise. Otherwise, other types of test as described below, applying non-constant stresses, are more relevant to the user, since they are more severe on the element's performance.

6.4 Single Element Test Approaches

Several approaches to single element tests can be devised, but here two main ones are discussed. Unlike the patch test, whose main aim is to prove that a given element type formulation admits constant stress fields, the single element tests have to consider higher stress variations in arbitrary geometries. This requires some care to ensure that all the nodes on the single element boundary have the known field behaviour suitably defined there, as either prescribed displacements or equivalent nodal loads. The higher applied stress variations can be linear, quadratic, cubic and so on. Some element types can give an exact response, the perfect convergence response, as defined in chapter 4. Thus, a quadratic displacement element gives correct results for a shear-loaded beam, which contains linear stress variations. Applied quadratic stresses on the other hand cannot be fully accommodated by a single QUAD8 element, all distortions giving large errors as seen in Tables 5.1 and 5.2, so the results will be poor. The QUAD9 element should perform better here. The lower order constant or bilinear stress elements would alone produce poor results and need mesh refinement for improvement. Higher order elements such as those with cubic displacements would give correct quadratic stresses. All these tests are useful in making comparisons of different distortion measures, such as varying aspect ratios, skewness, taper and any other geometric properties that might affect element performance. They are easy to set up and run, and yet can test all the above features, along with any inherent instabilities that characterise the element.

An obvious way to achieve this is to use the element as an interior patch of a known problem. The Continuum Region Element (CRE) method of Robinson [21,22] was developed for this, operating in either 2D or 3D. Consider as an example a 2D squat beam subjected to any loading where there is an analytical solution throughout the interior (figure 6.2) for all displacement and stress components. Hence, a single element of any shape can be constructed anywhere in this region. At every node, either prescribed displacements or nodal loads consistent with the assumed stress field have to be applied in each direction. The solution produces the unknown variables at each node, which can be checked for accuracy against the analytical solution. Sufficient fixings have to be made to prevent rigid body motion. More than one element can be used in the continuum, if required, and the whole continuum can be completely filled by one or more elements to still give a valid test.

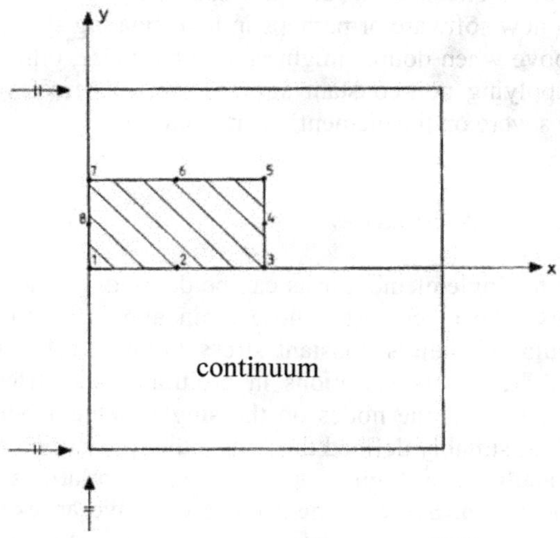

Figure 6.2 The Continuum Region Element Test

Another approach to single element testing is to use the element to represent the whole shape. Apart from basic tests used by code developers, these seem to have been first proposed as acceptable tests by Robinson [27]. The shape has to be simple, e.g. rectangular. The distortion shapes that can be studied are therefore limited, covering mainly variations in aspect ratio and skewness. Instead of using just one element, sometimes several elements are used together for good reasons, such as representing a rectangle by triangles, when pairs or fours would be needed. These cases are not intended to be a mesh refinement study.

6.5 Burrow's Tests

Burrows [20] used the CRE method to investigate shape sensitivity effects of the four common element types TRI3, TRI6, QUAD4 and QUAD8. The method allows each element to be distorted freely and yet have the correct applied loads and displacements at each node. Four types of shape distortion were considered, aspect ratio, skew, taper and edge curvature. The ranges of each were:

(a) aspect ratio; 1,2,3,4,8,
(b) skew; 0°, 15°, 30°, 45°, 60°, 70°,

(c) taper; 0%, 10%, 30%, 50%, 70%, 90%,
(d) edge curvature; -20%, -10%, 0, 10%, 20%.

The above values are defined in figure 6.3. Many runs were made for the different element types, varying these combinations. Some combinations were invalid in that over-distorted shapes resulted and det*J* became zero or negative. A particular beam geometry, defined in (x,y) coordinates, was subjected to, respectively, a constant direct stress σ_x, a constant moment and a linear moment, each converted to equivalent nodal loads. Minimal constraints were used. In each case, the various element shapes covered some suitable part of the beam, e.g. as shown in figure 6.2. Both displacement and stress errors were calculated at each node for each run and worst cases were tabulated. Stresses were calculated at both nodes and optimal points, along with extrapolation from the latter to the nodes.

Although the different combinations produced a large numbers of results, too voluminous to cite here, a few main conclusions are apparent. For constant direct or shear stress load cases, no errors were recorded, confirming that all the element types pass the patch test. For the constant moment load case, when expanded in polynomial form, the stress component σ_x varies linearly with height, y, and so the vertical displacement component is quadratic. For the linear moment case, σ_x is proportional to both y and xy, and the shear component is proportional to y^2. The displacement components are a degree higher and so offer taxing tests of the elements. From these last two load cases, the author found that:

- for TRI6 and QUAD8, aspect ratio and skew do not introduce errors as they are increased: det*J* is constant over the element in this case. Increasing the taper of QUAD8 does, however, increase errors. This taper is a group 3 distortion in figure 5.2,
- for QUAD4 and QUAD8, when the midside nodes are at the central positions, the only shape distortion parameter of significance is taper,
- errors start to grow as interior angles get close to 0° or 180°,
- TRI3 and TRI6 with midside nodes at the central positions only have errors when these interior angles get close to 0° or 180°. Again, det*J* is constant,
- curvatures introduce errors that depend on their magnitude and position in the element.

From these results, 5 specific benchmark tests have been defined. Four of the tests use constant moment loading and the other the linear moment loading. The tests are educational in that some of the distortions are well beyond what would be used in practice. However, they are useful tests of new software and help new users of established software to check out how well the error diagnostics work. To use them, the document by Burrows [20] should be consulted.

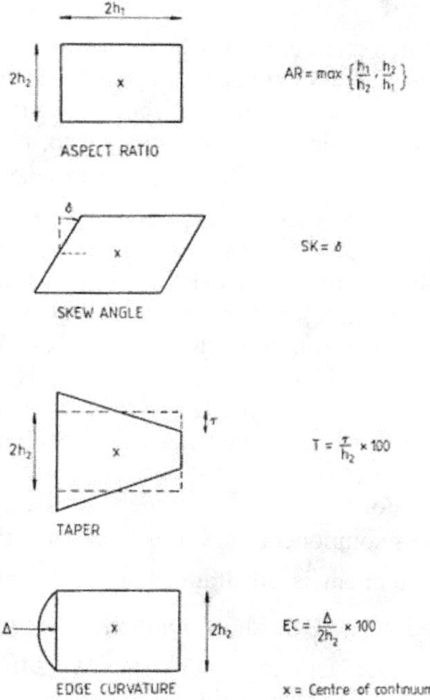

Figure 6.3 Element Distortion Definitions as used in the Burrows Tests

6.6 Some Simple Illustrative Tests

There are many ways of designing simple tests, with either one or just a few elements, and descriptions are given in many references in journals, books and software manuals. These can be consulted and rerun if necessary, to illustrate many of the points covered in this booklet. One particular such test is described below as giving good comparative behaviour of quadratic and cubic elements.

As an example of both the single element test and a convergence test, a beam subjected to quadratic shear (equivalent to a linear moment) on its right-hand edge and simply fixed on its left-hand edge has been proposed [27]. Under 2D plane stress conditions, varying aspect ratios of width to height are considered. The single element test results are shown in figure 6.4, and is similar to the test for zero energy modes in section 4.8.

For each run, the percentage error of the maximum tip displacement, v_1, is recorded for the single element types QUAD8, QUAD9 and QUAD12 and the 2 or 4 element representations of the triangular element types TRI6 and TRI0. Each element's curve is shown with an indication of the stiffness matrix integration rule used. Each element type was repeated with increasing aspect ratio, in the range 0.5 to 8.

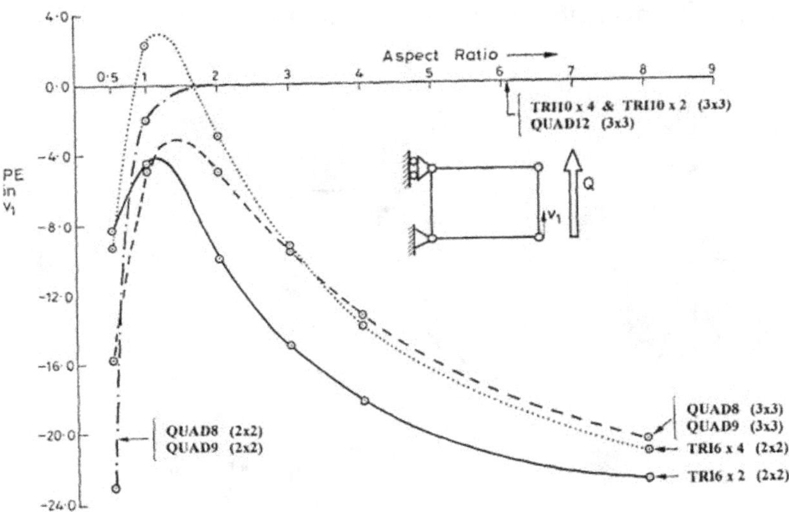

Figure 6.4 Percentage Error in Tip Displacement for Single Element Test

Figure 6.4 shows that there is no error in tip displacement v_1 for any aspect ratio when using the cubic elements QUAD12 and TRI10 (with both 2 and 4 triangles). Since the resulting stress field is quadratic, this ensures perfect convergence in these cubic displacement, quadratic or better stress, elements. All the other elements have quadratic displacements and therefore do not possess perfect convergence, so that the single element mesh represents the low end of the convergence curve. The error in v_1 is seen to increase rapidly to over 20% in all these elements except the combinations QUAD8 and QUAD9 with the reduced 2x2 integration rule, which are very accurate except when the aspect ratio is 0.5. As observed in section 4.8, these two results should contain zero energy modes, although the extent of hourglassing depends on the load case, which for this case is quite insignificant. This has to be seen as an interesting academic point, since QUAD8 and QUAD9 should not be used in earnest with the reduced integration rule in a single element application!

The errors in the higher aspect ratios are due to the high bilinear stress variation along the length-wise direction of the beam and the relatively coarse mesh in that direction. They are not due to distortion effects. The cubic elements model the situation well. For the quadratic elements, these errors would be expected to decrease with increasing numbers of elements along the beam. Hence, a convergence test has been conducted on this beam using meshes of 2, 4 and 8 elements along the beam, but retaining the maximum aspect ratio of 8 in each element, keeping only one element through the height. Figure 6.5 shows the rapid decrease in this error as the element numbers are increased.

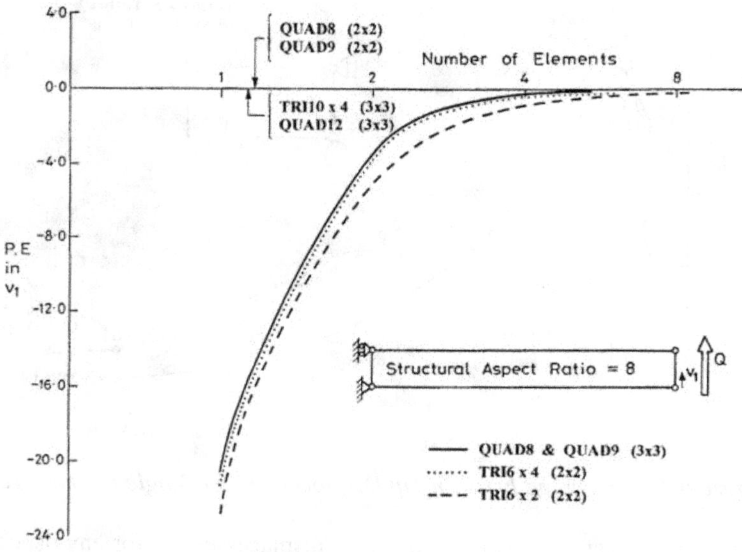

Figure 6.5 Percentage Error in Tip Displacement for Convergence Test

Thus, for practical meshes where refinements have been carefully considered by the user, reasonable aspect ratios should not introduce distortion errors, although it is still worth checking any software diagnostics on aspect ratio and any other measures, to ensure the elements are sympathetic to the expected local stress gradients. Although not included, the linear QUAD4 element would be a poor performer in this test.

6.7 Summary Comments

The above special tests have been described to illustrate various features of the more popular elements. The main intention here is to show the good and bad points of individual element behaviour in simple, yet sometimes very critical, tests. The user who is trying to gain experience in using these elements, or just gaining familiarisation with new software, is encouraged to repeat some or all of the above tests, or to seek out the reports of associated benchmarks and run some of those tests. In addition, if other element types are available, which have either not been covered in this booklet or which are still in the research or development stage, the above tests could be used to advantage to check on their performance.

7. Mesh Quality Indicators from Results

7.1 Introduction

One of the most important aspects of this booklet is the accuracy of results based on the chosen discretisation. This is a consequence of the approximate nature of the finite element method: the importance of choosing suitable element types and refinements has been central to earlier chapters. Although it is not possible to furnish a single value for the solution error from a given finite element run, some types of error in the various quantities calculated can be used for guidance on the quality of the chosen model.

Such output errors are highlighted in this chapter along with their usefulness. As has been alluded to earlier, secondary variables (stresses and strains) can be calculated in certain positions to show discontinuities across element boundaries, which give guidance to local mesh coarseness. These in turn are used to drive self-adaptive re-meshing algorithms. The stress errors are also a function of shape sensitivity, in that as an element becomes more distorted in some sense, nodal stresses can become more erroneous just because of the progressive distortion, and so some additional error guidance is available.

7.2 Calculation Errors

Formal error analysis is a complicated mathematical subject, and often only yields approximate conclusions or overall trends. Secondary variables are of particular relevance to most types of finite element analysis, since they are the only calculated quantities that can furnish useful errors guiding the accuracy of the solution. Stresses, and sometimes strains, have therefore been investigated for this.

Any particular error has to be carefully defined. Some relevant ones are described in [9]. The most usual type is, for any primary or secondary variable, the difference between the exact (usually unknown) value and the calculated value. Since the secondary variables like stress are very important in this context, the relevant stress error is

$$E_\sigma = \sigma - \hat{\sigma} \tag{7.1}$$

Here, σ is some calculated stress and $\hat{\sigma}$ is the exact value, at any reference point in the mesh. These symbols refer to any required component of the stress at any location in the mesh. Such pointwise errors will vary over the mesh depending on several factors. The most significant errors exist near singularities, such as at places where point loads or point boundary conditions are applied, or at sharp boundary features such as re-entrant corners and crack tips. Such local errors will be relatively large, even though the overall results may be accurate. If available, it is very useful to have access to these error measures all over the mesh and use them with care.

Another measure of error is the error norm, which is an integrated form of the above type of error. The energy norm $\|E_e\|$ is often used, and can be obtained from integrating the error of the product of stress and strain over the whole mesh:

$$\|E_e\| = \left[\int_V (\sigma - \hat{\sigma})^T [D]^{-1} (\sigma - \hat{\sigma}) dV\right]^{1/2} \quad (7.2)$$

where V is the total volume and [D] is the matrix of material constants. A corresponding, so-called L_2, norm of the stress error is:

$$\|E_\sigma\|_{L_2} = \left[\int_V (\sigma - \hat{\sigma})^T (\sigma - \hat{\sigma}) dV\right]^{1/2} \quad (7.3)$$

In finite element discretisations, these integrations are replaced by summations. The "root mean square" (RMS) value of the last error is the square root of the sum of the squares of each individual error such as given by equation (7.1), summed over the mesh. It therefore equals

$$\|E_\sigma\|_{RMS} = \left[\Sigma_i (\sigma - \hat{\sigma})_i^{\,2}\right]^{1/2} \quad (7.4)$$

The types of errors given by equations (7.2) to (7.4) are effectively smoothed over the mesh to produce a single value. They represent an averaged effect, and can be used for some purpose such as testing the state of convergence. The effects of severe but localised errors could therefore be missed. Because the error definition of equation (7.1) gives a single error value at each point for each component, a vector form is used to contain all of these values over all the reference points in the mesh. From this vector, it is particularly useful to examine the highest errors for guidance of behaviour at localised sites, as stated above, and use them alongside the averaged error measures.

Of course, the above error measures are only of use when there is some knowledge of the exact solution, and then they are somewhat academic since then the problem is solved! But some indications of approximations to the exact solution can be made and used instead of these exact values. These can be obtained from step changes in stress over element boundaries, as described later in this chapter.

7.3 Accuracy of Displacements

Displacements, the primary variable degrees of freedom, are directly calculated from the stiffness equations (2.13). These equations enforce equilibrium throughout the structure and so, conceptually, the displacements are most accurate at the nodes. At points in between nodes, they are given by the local shape functions and so depend on the surrounding nodal values. The approximate nature of these shape functions means that the accuracy of displacements away from nodes will be not therefore be quite as good as at the nodes. These primary variables are usually **conforming**, i.e. adjacent elements have the same values at all points along the common side or face. This means there is no more information that can be gained about their accuracy other than by convergence studies using mesh refinement.

7.4 Accuracy of Stresses

Strains, and therefore stresses, are derived from the displacements and so are not conforming, unless a special element formulation has been implemented (which is rare in commercial software). Thus, their accuracy varies more widely over each element. Along common sides or faces, it will be seen that they differ between the different elements meeting there. In the present context, it is sufficient to describe stresses - similar conclusions hold for strains, at least in elasticity. All the stress components calculated at a common node will be seen to differ when calculated in each of the adjacent elements, producing a lack of equilibrium, often known as a **stress jump**. As the mesh is refined, the magnitude of this jump should decrease.

The different ways of calculating stresses are described below. An important common feature is that, at any point and however calculated, the error in the stress component with the largest magnitude affects the other stress component errors by a similar magnitude, due to a Poisson's ratio effect. Thus, the smaller components absorb this error, which may therefore dominate their own errors. This gives a reduction in accuracy for the lesser magnitude components, and this reduction increases the smaller the value of the component. Only the stress component with the largest magnitude has a realistic error due to the state of convergence of the finite element analysis.

7.4.1 Nodal Stresses from Shape Functions

The **nodal stresses** are calculated using the element shapes functions in a differentiated form, as given by equations (2.8) and (2.11). Nodal stresses are frequently the main required outputs of the analysis, but unfortunately their accuracy is much less than the nodal displacements. Traditionally, the value at the node averaged over the adjacent elements has been used as the quoted nodal stress. Sometimes, this averaging process takes into account the inverse of the distance of each element's centroid from the subject node, to introduce a weighting effect.

Nodal averaging is not always applicable. For beam, plate and shell elements, the stresses are often in local coordinates which make averaging difficult or meaningless. Averaging should not be conducted over boundaries of different materials or thicknesses, along multi-planar junctions, and at locations of point loads or contraints. An expanded discussion of averaging is given in [7].

When they can be used, the averaged nodal stress values are often sufficiently accurate in practice, because the individual element values tend to oscillate above and below the correct value, particularly at nodes inside the structure's boundaries. However, such averaging for nodes on the boundary cannot be as accurate since much of the interpolation space is missing. Unfortunately, boundary nodes are usually where the key stress results are required. Some analyses require the calculated stresses to be of the highest accuracy since they may be required for on-going predictions in many engineering situations, such as fatigue and creep. This requires the mesh in such regions to be sufficiently fine. This will apply particularly to the highest stressed points of the mesh, where the above nodal averaging procedure would be at its least accurate.

7.4.2 Gauss Point Stresses

Within each element, the stresses are at their most accurate at the **optimal points**, or Barlow points, as defined in section 2.5. For many element types, these coincide with the reduced integration Gauss points. Such stresses are often referred to as **Gauss point stresses**. The stress and strain accuracy at these points is high, of the same order as the displacements, and so those components are said to be **superconvergent**. Although optimal point stresses should be used in all serious analyses, their locations are not usually convenient and so the nodal averaged stresses are still more widely used. When conducting non-linear analyses, the optimal point stresses are essential, since the on-going solution depends on accurate stress evaluation at each iteration. The averaged nodal values would not be good enough.

The triangular and tetrahedral elements do not have optimal points and so there is no superconvergence. However, their Gauss points are normally accepted for such use, to be consistent with other element types which do have optimal points.

7.4.3 Extrapolated Nodal Stress

A much improved **extrapolated nodal stress** can be calculated from the optimal point values, using a simple extrapolation of those values to the boundary. Figure 7.1 shows the principle behind this in the 2D quadratic quadrilateral QUAD8. The 4 optimal points are situated at $(\pm 1/\sqrt{3}, \pm 1/\sqrt{3})$ in theory space. The extrapolations are linear to the boundaries, in pairs, then linear again to the corner nodes or in to the midside nodes. This is consistent with the linear variation of the stresses throughout the element. The incomplete quadratic terms have to be ignored for this practical construction. At a common node, this process from all the surrounding elements will still give a stress jump, but this will be less than that calculated from the nodal averaging procedure of section 7.4.1, and with a more accurate average.

7.4.4 Superconvergent Patch Recovery

Some investigators have considered all the elements containing a given node and used a shape function approach to interpolate nodal values from all the optimal points contained in those elements. This is known as **superconvergent patch recovery** and is reported to give very accurate results [9]. Conceptually, it can be seen to be an improvement of the above extrapolation and effectively includes the incomplete quadratic terms as well as the nodal averaging. The interpolations would not be quite so accurate for boundary nodes.

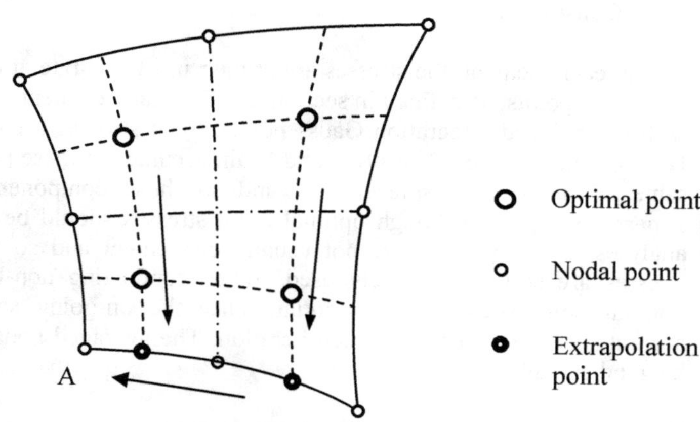

Figure 7.1 Nodal and Optimal Stress Points for QUAD8

7.5 Using Errors for Mesh Adaption

The above nodal stress jumps provide a practical advantage when used as measures of local mesh coarseness. In those areas where large nodal jumps occur, a finer mesh is indicated. This is the principle that drives self-adaptive remeshing algorithms as described in section 4.6. Indeed, this jump is the only indicator that the finite element method provides about its point-wise errors.

The general principle is based on the results from an initial elastic stress analysis. At any node on a typical element, suppose some component (including equivalent) of the nodal stress there has values as:

- σ_n if calculated at the node from the shape functions (section 7.4.1),

• σ^* if calculated by any improved scheme, such as extrapolating out from the optimal points (section 7.4.3),

Then the error measure $E = \left|\sigma^* - \sigma_n\right|$ indicates the error at that node and element. The way that this error is accumulated over the other nodes of the present element, and then over all elements, depends on the particular self-adaptive re-meshing software. The individual values of E need to be used to identify localised errors, rather than any smoothing over the element where such information would be diffused. An error estimate E_e for element "*e*" can thus be deduced, best as the largest value of E over the element.

Suppose that the required error that can be tolerated for the present analysis is $\overline{E}$, which has to be achieved in each element. For those elements where $E_e \le \overline{E}$, the error is small enough. For those elements where $E_e > \overline{E}$, there is a need to modify the element to reduce the error, or effectively to reduce the magnitude of the stress jumps, since there the stress gradients are relatively high. This provides the guidance for any h-convergence self-adaptive re-meshing algorithm. In some cases, the ratio $E_e / \overline{E}$ is used to indicate how much to change the element, and if $E_e \le \overline{E}$ it is possible to make the element larger, since the stress gradients are likely to be relatively low. A second analysis is then conducted and the stress errors assessed, again with a change of mesh if necessary. This process is repeated until the required level of accuracy is achieved. This leads to good uniform convergence since the errors tend to be roughly equal over the mesh.

The re-meshing algorithms are normally used with triangular and tetrahedral elements, since it is relatively easy for the re-meshing algorithm to fill 2D and 3D space with such elements. Quadrilateral and hexahedral elements give more accurate error estimates, because of the existence of optimal points inside them, but the automatic re-meshing is far more difficult since fitting space often causes adverse distortions.

7.6 Examples of Stress Calculation Errors

Any test problem can be studied to reveal errors in stresses, however calculated. Some interesting points are demonstrated using the examples of section 5.10. The simple square QUAD8 structure of figure 5.6 is considered with no distortions but with the relatively severe load case of parabolic applied shear and using reduced integration. The shear stress τ_{xy} varies quadratically with height, being constant in the x direction (figure 7.2). It is zero along the top and bottom edges and maximum along *y=0*. The stress σ_y is zero everywhere; σ_x is linear, zero along the right vertical edge and zero along $y=0$. Hence the greatest errors are expected in τ_{xy} along $y=0$.

The τ_{xy} shear stresses along $y=0$ are plotted against x in figure 7.2 for the 2x2 and 4x4 meshes of figure 5.6. The unaveraged nodal stresses are seen to oscillate, with maximum values of 121% and 105% of the theoretical value. Here, they are almost the same at common nodes, a feature of the symmetry of this test that is unusual in more complicated meshes. The results show that there is no perfect convergence, and that mesh refinement is required in the y direction. The quadratic stress implies that element size grading in that direction would produce an optimal mesh, but for the present, equally spaced elements are chosen. The 4x4 mesh shows much greater accuracy than the 2x2 mesh, and in each case the optimal point stresses lie within the oscillations.

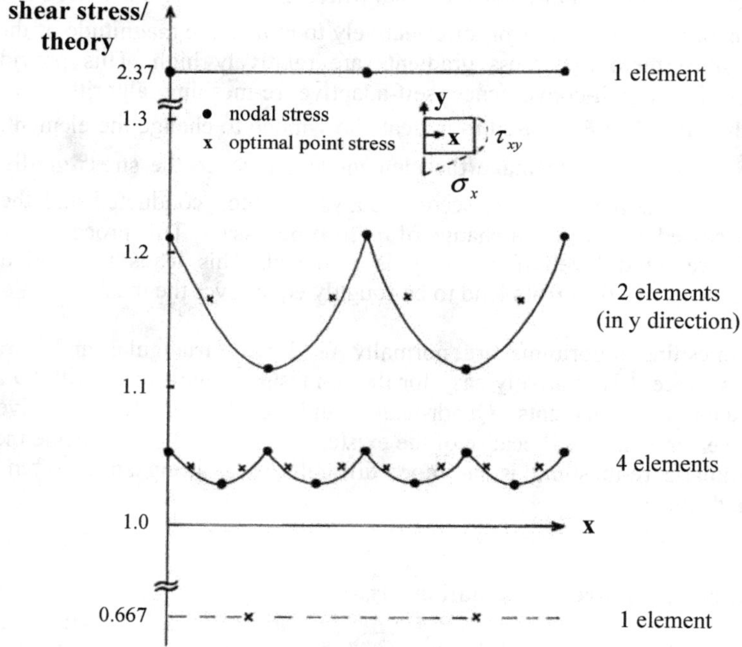

Figure 7.2 Shear Stress along $y=0$ for Different Element Numbers in y Direction

Also included in the figure is the result from the long beam test of figure 5.5 with the same load case. This is an even more severe test. Because there is only one element in the y direction, along the centre line $y=0$ there are no corner nodes, so the oscillations cannot be shown. Hence, there is only enough information to plot the nodal stresses as a straight line, at 237% of the theoretical value. Since there are only two optimal points in the y direction, the τ_{xy} stresses there have to be constant (therefore over the whole mesh). They give exact agreement with theory at those points, and cannot give any information about the maximum value along $y=0$, which is 1.5 times this constant value.

In all these meshes, the elements are capable of reproducing accurate σ_x stress values everywhere. This variation is linear, well within the capability of quadratic displacement elements, and high accuracy was obtained in the results. This highlights how stress components with lower magnitudes can retain high accuracy even when components with higher magnitudes are erroneous in unconverged meshes, provided these magnitude differences are not too large.

7.7 Errors due to Element Distortions

It has been seen in chapter 5 that element distortions can induce errors in the finite element solution. Table 5.1 shows how the total response errors are affected by increasing distortion and order of the strain field, both having a direct effect on the displacements, which are calculated from the stiffness equations. These displacement errors cannot be detected from the calculated results. The stresses subsequently calculated from these displacements reveal errors as indicated in Table 5.2, exacerbated by both increasing distortions and increasing order of the strain field. There is now a difference between the stress errors calculated at the optimal points and at other general points, typically the nodes. The optimal point stresses are more accurate than the nodal stresses, and so, unlike displacements, some degree of error magnitude can be ascertained. At this point, it should be clarified that the nodal stresses are derived from the element shapes functions at each node in a differentiated form as given by equations (2.8) and (2.11), section 7.4.1. We are concerned with behaviour inside any one element so no nodal averaging has yet taken place.

7.7.1 Stress Error Measures

To proceed, consider the quadratic quadrilateral QUAD8 element, whose optimal points and extrapolation method are shown in figure 7.1. Assume that an elastic stress analysis has been conducted. At any node on some such element, suppose some component (including equivalent) of the nodal stress there has values:

- σ_n if calculated at the node from the shape functions (section 7.4.1),
- σ_{gn} if calculated by extrapolating from the optimal points (section 7.4.3),
- σ_{true} is the exact value.

Consider a test where this element only experiences a linear stress field, such as in a simple bending problem, or for a small element inside a higher order of strain loading. The optimal point stresses and hence σ_{gn} will be exact. If the element is not distorted, the nodal stresses σ_n will also be exact. It can readily be shown that as the element is progressively distorted, then the value σ_n becomes erroneous. This can be measured and compared to an *a priori* estimate of the distortion such as the maximum/minimum ratio of det*J* over the element.

Suppose now that a quadratic or higher stress field applies across the element. The optimal point stresses are now in error and hence σ_{gn} will also show an error of $E_{gn} = |\sigma_{gn} - \sigma_{true}|$, and the nodal stress σ_n will show an error $E_n = |\sigma_n - \sigma_{true}|$, both due to the model discretisation. Absolute values of the differences are sufficient here. When shape distortions become significant, however, experience shows that the nodal stresses develop greater errors than the optimal point stresses, i.e. the error E_n dominates the error E_{gn}. This is supported by the quadratic strain order of Table 5.2, where the two errors are of order O(1) and O(*d*) respectively.

Thus, although the true stress and therefore the discretisational stress errors are in general not known, the difference $E_s = |\sigma_{gn} - \sigma_n|$ is of importance for predicting the inaccuracies due to shape distortion, since this difference reflects the error E_n in the nodal stress. This is because, as the degree of distortion increases, E_n dominates the difference in the discretisation errors $|E_{gn} - E_n|$. For linear stress fields, $|E_{gn} - E_n|$ is negligible and so $|\sigma_{gn} - \sigma_n|$ gives a good prediction of the actual nodal stress error. For quadratic stress fields, $|E_{gn} - E_n|$ is unknown but is dominated by the E_n nodal stress error, the more so the greater the distortion.

Hence, the stress difference $E_s = |\sigma_{gn} - \sigma_n|$ is seen to give a good estimate of the nodal stress error resulting from element distortion. Although the present argument is only for the QUAD8 element and for linear and quadratic stress or strain fields, the conclusions should also hold for other element types, particularly the closely-related QUAD9, HEX20 and HEX27 elements. They should also apply when higher stress gradients exist.

As an example of other element types, consider the constant stress triangle TRI3, where the optimal point lies at the centroid. Whatever the applied stress field, the nodal stresses on one element have to be the same as the optimal point stresses, since the stress is constant over the element. No deviation is allowed for these stresses, although being a linear triangle, no distortion below group 5 of chapter 5 is possible.

7.8 Examples of Errors due to Element Distortions

In test problems where the true stress solution is known (so that, at the node of interest, σ_{true} is known) a stress error diagram can be drawn to plot the stress error measure $E_s = |\sigma_{gn} - \sigma_n|$ against the known error in σ_n (i.e. $E_n = |\sigma_n - \sigma_{true}|$). If the predictions are good, then the point will lie close to the 45° line. As the distortions increase, so do the errors, therefore the corresponding points will lie further away from the origin. A measure of the distortion can be gauged as a single number using the maximum/minimum ratio of det*J* over the element. Such a scheme is adopted in the following example.

The square beam described in section 5.10.2 is presented to study these predictions against known analytical solutions. The same two load cases are considered, the constant applied moment and parabolic shear loading. Using the finest 4x4 mesh of QUAD8 elements, both load cases render excellent agreement with the analytical solutions using both complete and reduced integration when all the elements are square. Progressive distortions were introduced in the top left-hand four elements, where the stresses are highest, using internal node movements so as not to disturb the external boundary shape, in a variety of ways including those indicated in figure 5.6.

For the constant moment loading, the horizontal stress component σ_x varies linearly with the height, *y*, and the other components are zero. This case was restricted to a number of group 3 distortions of figure 5.2. This involved moving corner nodes in the top left element, retaining the midside nodes in their central positions. With even quite large distortions the errors in the nodal stresses remained very small for most distorted shapes. However, in one case, where the corner node opposite to the top left-hand corner node in element 1 was progressively moved towards the element centroid, large errors in that nodal stress appeared, increasing as the centroid was approached. Errors of 19.4% appeared in the stress error measure $E_s = |\sigma_{gn} - \sigma_n|$, and 20.0% in $E_n = |\sigma_n - \sigma_{true}|$ corresponding to a det*J* ratio of 10 (the corner node was very close to the undistorted centroid position). This would therefore give a point very close to the 45° line in a stress error diagram. Such a diagram is not shown for this load case since few errors of any significance were recorded.

For the parabolic shear load stress case, many distortions were studied, some being quite severe and even extreme at, for instance, the CC′ positions where det*J* becomes zero. These covered groups 3, 4 and 5 of figure 5.2. In this case the direct stress components are linear and the shear stress is quadratic. The net behaviour is therefore quadratic and so there will be significant non-zero values for the stress error measures $E_s = |\sigma_{gn} - \sigma_n|$ and $E_n = |\sigma_n - \sigma_{true}|$ when distortions are introduced into the mesh. Since analytical results are known for this load case, a stress error diagram can be constructed. Such a diagram is figure 7.3 for many points representing the different distortions. Although details of distortion types are not shown, each point is distinguished by the det*J* ratio pertaining to that particular distortion, and is calculated from a particular stress component at a key node in the distorted region.

Figure 7.3 shows that there is indeed a tendency for the points to lie on or close to the 45° line, supporting the above assertions. There is also a tendency for larger error magnitudes to occur when larger det*J* ratios exist. Thus, a high value of $E_s = |\sigma_{gn} - \sigma_n|$ predicts a high nodal stress error, which in turn is a direct result of mesh over-distortion. Even when points are not close to the 45° line, if the error is large enough, the indication is that a significant predicted error E_s will occur when there is a significant nodal stress error. Since E_s is readily calculated, a useful *a posteriori* guide to the state of the mesh and its distortions is available. The *a priori* det*J* ratio is generally not quite such a reliable guide.

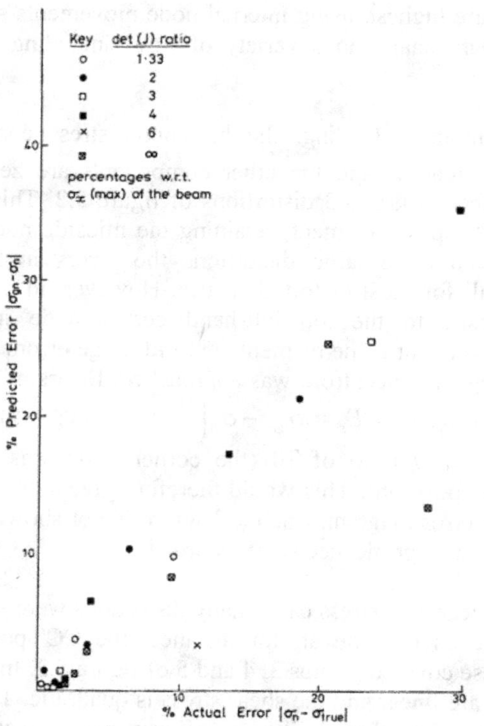

Figure 7.3 Predicted versus Known Stress Errors for Beam under Linear Applied Stress

7.9 Summary Comments

The earlier chapters of this booklet have been concerned with the importance of choosing suitable element types and refinements for designing a suitable mesh. Advantage can be taken of any knowledge of expected behaviour to achieve fairly optimal meshes, although of course there is no way of knowing *a priori* the true results for non-trivial problems. However, useful information has been shown to be available from the results of the finite element analysis in this chapter. For isoparametric elements, errors calculated from the stress results, deduced from the lack of stress equilibrium across element boundaries, have been described. They give guidance to local mesh coarseness and can be used to drive self-adaptive re-meshing algorithms. The errors have also been shown to depend on shape sensitivity, which can give a guide to detrimental distortion effects. These various issues have been illustrated by examples that can be potentially adaptable to other situations, such as when conducting important new problems for which the correct results are not known, even to approximate values.

8. Concluding Remarks

This "How To..." booklet has discussed the effective use of the more common element types in designing suitable finite element meshes for solving real engineering problems. Such elements include the isoparametric element families that are used in modelling 2D and 3D geometries, and various special shapes such as beams, plates and shells.

When designing a finite element discretisation of a given structure, the user has to be aware of two important considerations. The first concerns the overall modelling strategy, which includes how much of the given structure needs modelling, including any advantages of symmetry or anti-symmetry. It also includes the geometric shape involved, what detailed features are to be included, and what aspects of the boundary and load conditions are relevant. Such topics have been described in chapter 4 and discussed in greater detail in other "How To..." texts, e.g. [5]. The second consideration, which follows, is the main concern of this booklet, namely on the usage of the actual element types. The relevant aspects that the user needs to be aware of include:

- suitable choice of element type: order of displacement variation, shape (triangular, quadrilateral, tetrahedral, hexahedral, etc.),

- use of numerical integration rule, complete or reduced, as relevant,

- issues concerning evaluation of stresses (optimal points, etc.),

- element spacing: the user may choose a carefully-designed mesh with element sizes smaller in areas of high stress gradients and larger in areas of lower gradients, or with higher performance elements using advantages such as reduced integration. Alternatively, the user may choose to use a large number of elements over the model, fine everywhere, but creating larger computer models, data storage and run times. Such meshes are often easily generated by automatic schemes. If insight into the expected stress behaviour exists, then the map analogy may be used to design efficient meshes for the first category. The elements would thus simulate the stress behaviour in a sympathetic way, with good gradings, shapes and optimal numbers for run-time efficiency,

- allowable distortions of elements: carefully-designed meshes of high performance elements, as described above, may well require distorted shapes of quadrilaterals or hexahedra. Such distortions should be sympathetic with the expected stress gradients, as described in chapter 4, and would then tend to enhance rather than diminish local accuracy. Alternatively, the user may choose to use larger numbers of triangular or tetrahedral elements over the model, again creating larger computer models, data storage and run times. Such meshes are often easily generated by automatic schemes.

In the last two aspects, concerning element spacing and element distortion, the distinction is made between using carefully designed meshes of higher performance elements using advantages such as reduced integration, or of using greater numbers of elements over the model, incurring larger computer models, data storage and run times. The choice can depend on the conditions and importance of the analysis. For instance, if the user has to work quickly, to produce results on a one-off basis, then the second option of using large numbers of easily-generated elements, probably triangular or tetrahedral, but accepting the larger computer models, may well be preferred. On the other hand, if a more deliberate approach can be taken, the experienced user could produce a mesh with carefully chosen element spacings and distortions, and use element types that can take advantage of reduced integration to produce good optimal point stresses. This is useful when the analysis is likely to be repeated, such as in a series of self-similar design studies, or when non-linear analyses are required.

Any kind of non-linear analysis may involve a series of computer runs, each involving many iterations of the solution of the basic stiffness equations (or their equivalent). The desire for run-time cost efficiency and suitable accuracy implies that an efficient mesh is highly desirable. Since the repetitive non-linear processes require a continuous evaluation of the stress state, optimal point stresses are essential. Much of the discussion in this booklet has assumed that the behaviour of stress and strain is the same, as in elasticity. However, in non-linear situations, the constitutive laws render one varying locally more than the other. These comments would therefore apply to whichever varies the most. In many cases, for instance in plasticity, the stress variation is severely restricted by some imposed yield condition inside the plastic zones, whereas the strain variation there can become particularly complex. A similar unequal response may occur in geometric non-linearities, where the strain variation can become very large. Some non-linear solutions are difficult to formulate for elements of order higher than unity, and many commercial codes recommend the use of linear elements in situations such as contact analysis.

Of great assistance to the discussion on element suitability is the availability of error measures, as described in chapter 7. Here, the lack of equilibrium across element boundaries is measured from the calculated nodal stresses. Optimal point stresses cannot be used for this since they are evaluated inside each element. The resulting "stress jumps" indicate whether the mesh is too coarse, or even too fine, at that particular boundary. This information gives a useful assessment of accuracy over the whole mesh. It is also used to drive the following loop in self-adaptive re-meshing algorithms. These are based on generating triangular meshes since they are easier to adapt automatically. Software for quadrilaterals is much less common, as is software for adaptive meshing in 3D.

If such automatic schemes are to be used, it is sufficient to generate the initial mesh using a relatively large number of triangles and then let the algorithms do the rest. The user provides an indicator of the required error to be tolerated, from which an optimal mesh is eventually produced, containing elements of suitable size and spacing and with a uniform prescribed stress jump error everywhere.

9. References

1 Baguley, D. and Hose, D.R., Why Do Finite Element Analysis, NAFEMS Order No. HT0, 1994.

2 Baguley, D. and Hose, D.R., How To Plan a Finite Element Analysis, NAFEMS Order No. HT6, 1994.

3 Baguley, D. and Hose, D.R., How To Get Started with Finite Elements, NAFEMS Order No. HT5, 1994.

4 Baguley, D. and Hose, D.R., How To Choose a Finite Element Pre- and Post-Processor, NAFEMS Order No. HT4, 1994.

5 Baguley, D. and Hose, D.R., How To Model with Finite Elements, NAFEMS Order No. HT7, 1997.

6 Marks, L., Tips and Workabouts for CAD Generated Models, NAFEMS Order No. HT13, 1999.

7 Baguley, D. and Hose, D.R., How To Interpret Finite Element Results, NAFEMS Order No. HT8, 1997.

8 Cook, R.D., Malkus, D.S. and Plesha, M.E., Concepts and Applications of Finite Element Analysis, 3rd Edition, John Wiley and Sons, 1989.

9 Zienkiewicz, O.C. and Taylor, R.L., The Finite Element Method, 5th Edition, Butterworth-Heinemann, 2000.

10 Adams, V. and Askenazi, A., Building Better Products with Finite Element Analysis, OnWord Press, Santa Fe, New Mexico, 1999.

11 MacNeal, R.H., Finite Elements: Their Design and Performance, Dekker, 1994.

12 Barlow, J., Optimal Stress Locations in Finite Element Models, Int. J.Num. Meths. Engng., 10, 243-251, 1976.

13 Barlow, J., All You Ever Wanted to Know About FE Analysis and Were Too Afraid to Ask, Lecture Notes for Cranfield University Course, 1990.

14 Roark, R.J. and Young, W.C., Formulas for Stress and Strain, McGraw-Hill, 1986.

15 MacNeal, R. H. and Harder, R.L., A Proposed Standard Set of Problems to Test Finite Element Accuracy, Finite Elements in Design, 1, 3-20, 1985.

16 Honkala, K.A., Adequate Mesh Refinement for Accurate Stresses, in NAFEMS Advanced Workbook of Examples (Volume 1), NAFEMS Order No. R0078, 2001: also summarised in NAFEMS Benchmark, January 2000.

17 Robinson, J., An Introduction to Hierarchical Displacement Elements and the Adaptive Technique, NAFEMS Order No. P13, 1986.

18 Smart, J., Back to Basics: Element Shape Distortion, NAFEMS Benchmark, April 1999.

19 Hellen, T.K., How To Undertake Fracture Mechanics Analysis with Finite Elements, NAFEMS Order No. HT18, 2001.

20 Burrows, D.J. and Enderby, L., Shape Measuring Criteria and the Establishment of Benchmark Tests for Single Membrane Elements (Summary), NAFEMS Order No. R0012, 1993.

21 Robinson, J., Basic and Shape Sensitivity Tests for Membrane and Plate Bending Finite Elements, NAFEMS Order No. P01, 1985.

22 Robinson, J. CRE Method of Element Testing and the Jacobian Shape Parameters, Engng. Comps, 4, 113-118, 1987.

23 Robinson, J., Distortion Measures for Quadrilaterals with Curved Boundaries, Finite Elements In Analysis and Design, 4, 115-131, 1988.

24 Barlow, J., More on Optimal Stress Points – Reduced Integration, Element Distortions and Error Estimation, Int. J.Num. Meths. Engng., 28, 1487-1504, 1989.

25 Irons, B.M. and Ahmad, S., Techniques of Finite Elements, Ellis Horwood Ltd., 1980.

26 NAFEMS Publications Catalogue, NAFEMS, regularly updated.

27 Robinson, J., Element Evaluation – A Set of Assessment Points and Standard Tests, Second World Congress on Finite Element Methods, Bournmouth, England, October 1978.

www.ingramcontent.com/pod-product-compliance
Lightning Source LLC
Chambersburg PA
CBHW052016230326
41598CB00078B/3491

* 9 7 8 1 9 1 0 6 4 3 2 2 8 *